BIG DATA TRANSPORTATION SYSTEMS

Guanghui Zhao

Guizhou University of Finance and Economics, China &

Transportation Think Tank Information Technology Research Institute of MOT, China

Gusheng Zhu

Qujing Normal University, China

Published by

World Scientific Publishing Co. Pte. Ltd.
5 Toh Tuck Link, Singapore 596224
USA office: 27 Warren Street, Suite 401-402, Hackensack, NJ 07601
UK office: 57 Shelton Street, Covent Garden, London WC2H 9HE

Library of Congress Cataloging-in-Publication Data
Names: Zhao, Guanghui, 1976– author. | Zhu, Gusheng, 1963– author.
Title: Big data transportation systems / Guanghui Zhao, The Guizhou Transportation and Economy Technology Institute, China, Gusheng Zhu, Qujing Normal University, China.
Description: Hackensack, NJ : World Scientific, [2021] |
Includes bibliographical references and index.
Identifiers: LCCN 2021009807 | ISBN 9789811235993 (hardcover) |
ISBN 9789811236006 (ebook for institutions) | ISBN 9789811236013 (ebook for individuals)
Subjects: LCSH: Intelligent transportation systems. | Big data.
Classification: LCC TE228.3 .Z43 2021 | DDC 388.3/12028557--dc23
LC record available at https://lccn.loc.gov/2021009807

British Library Cataloguing-in-Publication Data
A catalogue record for this book is available from the British Library.

大数据交通: 从认知升级到应用实例
Originally published in Chinese by China Machine Press

For any available supplementary material, please visit
https://www.worldscientific.com/worldscibooks/10.1142/12250#t=suppl

Desk Editors: Jayanthi Muthuswamy/Yu Shan Tay

Typeset by Stallion Press
Email: enquiries@stallionpress.com

Printed in Singapore

Foreword

On May 10, 2017, my student Dr. Zhao Guanghui and I discussed issues related to the development of intelligent traffic at AIMS, Hangzhou. Later, he planned to write something about traffic big data. I told him that the University of Michigan has the world's leading intelligent traffic laboratory. They spent a few years trying to optimize several traffic lights next to the University of Michigan. Then, they came to a big city of China to start experimenting on traffic big data by managing traffic lights with big data. With the help of the traditional camera and the induction coil, the light was switched every 90 seconds through man-made operation. By analyzing the congestion data, which was sent to the monitoring room, the traffic lights were adjusted through artificial intelligence. After two months, the city's congestion eased 20% during rush hours. In the future, more cities in the world will optimize their traffic lights through big data, which will help maximize the use of the limited resources of road and save a lot of time and traffic resources. What is more, there are 150–200 million privately owned cars in China. These cars are idle in the parking lot about 95% of the time every day. They take up much space, waste resources, and cause congestion. With big data, these idle resources can be mobilized to serve the people who need to take a ride. How much contribution the practice would make!

After reading the manuscripts of the book written by Dr. Zhao Guanghui and Professor Zhu Gusheng, I deeply believe that the big data technology has promoted the development of China's traffic. The examples include collection of traffic big data sources, application of traffic big data platforms, the pain points, cases and business models of traffic logistics, public transportation, and traffic safety, operation experience, and the current status, strategies and actions of traffic big data in other countries. Dr. Zhao proposed that traffic big data has changed the travel patterns of Chinese society, shifting from "people trying to find a car" in the past to "making arrangements for cars to find people who want to take a ride" now. By analyzing data on rainy days, sunny days, rush hours, normal hours, weekdays, and holidays, the future travel conditions and number of passengers in specific places under these circumstances can be predicted. The accuracy rate has now reached 85–90%. Decisions can be made 15 minutes earlier to send the cars to designated places. With the allocation, route planning, and positioning 15 minutes in advance, online car-hailing brings much convenience to people, which is unimaginable before. Now people get quite used to it.

The effort to match passengers and cars is not the only application. The effort to match people and materials has also changed greatly because of big data. At 24:00 on November 11, 2017, the sales of Tmall on that day reached 168.2 billion yuan. Among them, online shopping contributed 90%, with sellers and buyers covering 222 countries and regions. Before November 11, sellers predicted by big data that express parcels would increase by more than 30% that of 2016, and the big data predicted that the number would reach 140 million on November 11, and more than 1.7 million couriers would be involved in the logistics. By enhancing collaboration and data sharing to improve logistics services, the delivery bills had covered more than 80% of the whole industry. It was the first time that big data forecast cover more than half of the express parcels in the industry, and the efficiency was improved through the connection of big data and social collaboration. On the basis of the upgrade of logistics early warning radar, the big data platform has fully elevated from the time interval forecast to the real-time forecast, covering more than 50% of China's express parcels. The algorithm optimization enabled big data of more dimensions to be used. As a result, real-time forecast of number of parcels delivered every day was provided, with the accuracy rate exceeding 90%. These data can help express partners to deploy capacity, and companies can prepare capacity in advance based on the forecast data. Therefore, efforts were made to increase 4,000 temporary workers and 5,240 trucks, and increase the tonnage of plane leases in advance. The "big data distribution route" can automatically calculate the distribution path and allocate the parcel to the outlet closest to the buyer based on the information on the delivery bill. More than 50% of parcels were delivered to buyers faster through big data distribution route. Sellers were giving forecast of all kinds of best-selling goods by using big data, and managing inventory at warehouses across the whole country, so that the parcels could be as close to buyers as possible. By using big data to forecast the volume of parcels, the goods were able to reach buyers as early as possible. The pattern shifted from "buyers waiting for the parcels" to "parcels finding buyers".

Big Data Transportation Systems brings us a brand new perspective to observe and make use of big data, thanks to Dr. Zhao Guanghui's hard work after returning from the University of Michigan and Professor Zhu Gusheng's continued attention to China's traffic. I wrote the preface to *Internet + Traffic*, a book written by them before. This book *Big Data Transportation Systems* provides a depiction of the new stage of traffic development. Against the backdrop of China's strategy to develop itself into a transportation power, such forward-looking study, the scholars' interaction with the workers in the transportation field, and scholars' contribution to the development of world transportation became all the more necessary. Everyone must have the purpose of life, because your existence has an impact on society and the people around you. Everyone can make a difference to the world no matter what role s/he is playing. I hope that Dr. Zhao will continue to work hard, stay diligent, grasp the major technological changes in the world, follow the world's trend, provide more research findings for the transportation power dream with his international vision, broad mind and extraordinary wisdom and deliver more works to readers.

I believe, thanks to the wisdom of China's talents in the field of transportation and the contribution of talents from all over the world, China's traffic big data will definitely become more advanced in the future. China will come closer to its goal of becoming a world transportation power, and the prospects of China's traffic development by virtue of big data will also be brighter.

Ni Jun
University of Michigan, US
December 12, 2017

Foreword

The big data is a popular area that has attracted worldwide attention, as it will profoundly change the way people work and live. Its huge potential application has been fully reflected in many fields. By applying big data technology, Caoniao Logistics has established China's smart logistics backbone network supported by "Sky Network + Ground Network + People Network"; toutiao.com actively constructs intelligent recommendation algorithms with personalized design, wide coverage, and high input–output ratio; Amazon is focusing on the development of intelligent, Internet-based, and automated smart supply chain management systems. The application of big data in traffic is also keenly anticipated. Traffic big data refers to the collection of data featuring large-scale, multiple sources, various types, complex structures, and low value densities. By exploring its connection and common rules by using big data technology and analytical methods, we can help provide vigorous data support for the management decisions of relevant departments, accelerating the successful implementation of intelligent traffic and smart cities.

Although China's transportation industry just started not long ago, it has made remarkable progress by drawing on the advantages as a late comer. As a country with a vast territory, complex topographical and climatic conditions and huge geographical differences, China has accumulated rich experience in the research, design, testing, construction, maintenance, and management of infrastructure projects after years of practice. We have overcome worldwide problems in building highways and railways on plateau frozen soil, expansive soil and in deserts, and our high-speed railways have become a new example of "Made in China" and have been introduced to other countries. However, as a basic, forerunner, and service-oriented industry supporting the development of the national economy, the transportation industry is a large and complex system. Catching up with transportation powers such as Japan, Europe, and the United States is no easy feat. To develop the transportation industry, pouring capital investment is far from enough. We not only need to provide sufficient talents, fund, and technology but also need to formulate scientific and rational strategic planning, effective implementation programs, and comprehensive long-term guarantee mechanisms at the top level.

There is no doubt that governments at all levels in China have attached great importance to the development of the transportation industry. Thanks to their hard work, the mileage of roads and railways has grown rapidly; the construction of public transportation system and subways has been going on in an orderly manner; and the emerging model of "Internet + Traffic" such as shared car rental and shared bicycles have also been flourishing under the support and guidance of the local governments. However, China's transportation industry, as a whole, is still at the initial stage of development. Unbalanced regional development, lack of industrial inter-connectivity, and lag in information technology have caused many problems, such as traffic congestion, frequent traffic accidents, low transportation efficiency, and high costs. These problems have brought inconvenience to people's work and lives, and restrained the steady and sound growth of China's economy. The emergence of big data traffic provides an effective solution to the aforementioned problems.

Developed countries such as the United States and the United Kingdom have conducted in-depth research and practice on the application of big data in traffic, achieving good results in solving traffic congestion, emergency response in severe weather and road condition analysis and prediction. Governments at all levels in China are working hard to catch up, as they have actively explored the application in Beijing, Shanghai, and Hangzhou, setting a good example for the whole country. In fact, China has invested a considerable amount of resources in the process of promoting traffic big data. However, due to the lack of overall strategic planning, problems such as repeated development and low resource utilization are particularly serious. Meanwhile, talents in big data are quite scarce. Even in developed countries, talents in this regard are scarce resources, therefore, the research on the theory and application of big data is still quite limited. Interdisciplinary talents in China who can apply big data technology and analytical methods to transportation are even more scarce.

What is more, the problem of information island is particularly serious. The ability to collect, analyze, and apply massive traffic data from multiple sources efficiently and cost-effectively is the foundation and precondition for advancing the implementation of traffic big data, but information systems among different regions, levels, and departments of China have yet to align seamlessly, hindering the spreading and sharing of traffic data. Therefore, breaking the information islands among different regions, levels, and departments has become an important task in the development of traffic big data in China. This book provides a systematic and in-depth analysis of the macro-background of the rise of traffic big data, and the development status, pain points, priorities, strategic planning, and implementation programs of traffic big data. It recommends a way to develop traffic big data with Chinese characteristics. The book's systematic review of the development of China's traffic big data is impressive. In particular, while summarizing traffic big data cases of transportation powers such as the United States and the United Kingdom, the book also gives detailed analysis of the exploration and practice in this area by Beijing, Shanghai, Guizhou, and Yunnan. It provides valuable experience for decision-makers, managers, governments at all levels, technical service providers and equipment manufacturers. The development of traffic big data will not only accelerate the in-depth application of the next-generation information technology such as big data, cloud computing, sensors and artificial intelligence in the transportation area, but more importantly, it will bring a new way of thinking and development philosophy to the transportation industry.

Compared with Europe and the United States, China still lags behind in the traffic big data software and hardware such as technology, equipment, and information systems, but we can catch up if we learn from their experience. However, if we lag behind in our way of thinking and development philosophy, we will remain imitators and followers forever, and the dream of becoming a transportation power will never come true. Therefore, in the future, China should seize the important opportunities brought about by traffic big data, innovate our ways of thinking and development philosophy, introduce advanced technologies, equipment, and management models, and catch up with them through developing traffic big data, so as to turn ourselves from a big transportation country into a transportation power.

Zhao Guanghui is Professor Ni Jun's student. As a young talent in science and technology recognized by the Ministry of Transport, he pursues rigorous and precise study of big data, and he is good at innovation and exploration. I am so delighted to write a foreword his book *Big Data Transportation Systems*. I headed a team in 2014 to the M. Wu Manufacturing Research Center of the University of Michigan to study the US manufacturing. At that time, Professor Ni Jun happened to be in China. Dr. Zhao told us at the seminar that he used to work at the Ministry of Transport, and came to the University of Michigan to study big data manufacturing and traffic big data. It was a very rewarding visit. Four years have passed, but I still remember vividly taking photo with Dr. Zhao on the campus of the University of Michigan. In the past four years, he wrote a series of books on traffic big data, some of which have been published overseas. He is always tireless and diligent, and young scholars can learn from him.

You Zheng
Tsinghua Campus, China
April 13, 2018

Preface

In today's world of rapidly expanding information, the data have become a valuable resource for the development of human society. The Internet age, in the form of e-commerce, e-office, e-government, etc., has brought human data communication to the PB level, and in most scenarios of our lives, the big data have become a way and a synonym to describe the large amount of unstructured and semi-structured data.

In particular, the value and role of big data in the field of transportation, which is closely related to our lives, is constantly highlighted. The vast amount of data and information in the transportation sector, if not properly mined and used, will only be wasted as valuable resources. Moreover, the goal of intelligent, personalized, and self-service transportation will be an empty word.

In a sense, big data transportation is the intelligent transportation and self-service transportation. With the support of big data, transportation will be more reasonable and more efficient.

Since the rapid development of the field of intelligent transportation in 2005, big data has been highly valued by the government as a goal to "ensure safety, enhance efficiency, improve the environment and save energy". Many of the technologies in the field of transportation are at the international leading level. With the increasing popularity of big data, big data transportation and intelligent transportation are more and more pervasive. Related products and services are also constantly entering people's lives.

For example, shared bicycle, Didi taxi, carpooling, self-service travel, Gaode map, and intelligent logistics are intelligent products and services which show vigorous vitality under the support of big data, and transportation has become more efficient and environmentally friendly.

The development of big data transportation is considered of great importance not only by China but also is an important topic all over the world. However, Japan, the United States and Europe, which developed earlier, have their own unique technology and experience. These techniques and experiences are worth studying and learning from in order to better develop our big data transportation. We may not resolve our own defects by accepting other's good suggestions, but it is possible to turn them into our own use. In the development and research of big data transportation, we have to learn from others, we have to be good at the application and imitating things can never be independent.

Throughout the book, the discussion and research are explained in Parts I and II, and the order of cognition first and then practice also fits the cognitive habits of the readers. The book cites a large number of domestic and foreign cases, including relevant government action cases, as well as cases of exploration by enterprises in an attempt to present a comprehensive picture of the present and future of the development of big data transportation.

The development of big data transportation has not been smooth sailing. Although much has been achieved, there are still various problems. For example, the development of big data transportation in China is relatively chaotic and lacks in the formation of a unified layout. The administrative regions are somewhat fragmented and lacks in a unified standard, which is not conducive to the future development of big data transportation, and data without standards cannot be used as a whole. If transportation data is shared in the future, it is likely to be costly and time-consuming to harmonize the formats and standards.

Besides, because the big data transportation started off late in China, at the level of data transportation sharing, there are no big platforms to realize the opening and sharing of data resources. These aspects have a long way to go compared to foreign countries. Without advance planning and research, various obstacles may arise in the future.

Overall, the development of big data transportation in China has been very successful and the future is extremely bright. On the basis of development in accordance with national conditions, we should learn more from foreign experience, plan ahead, and take advantage of the early layout. Accurately grasping the advantages of big data in the field of intelligent transportation plays a very important role in improving transportation efficiency, solving traffic congestion, ensuring transportation safety, reducing environmental pollution, and improving China's traffic conditions at a new height and starting point.

In the future, we look forward to more convenient and easy transportation. No matter the government or the enterprise, they can explore better modes in big data transportation and intelligent transportation, so as to realize a bright future with smooth transport and communication, friendly environment, and convenient transportation.

The book was written in a hurry, which inevitably leads to various mistakes. I am grateful for all readers correcting them enthusiastically!

About the Authors

Guanghui Zhao (1976—) is researcher, professor, and bachelor of law at Wuhan University. He has completed his master's degree in administration from Central South University of Finance and Economics; master's in economics from Central Agricultural University; Ph.D in management science and engineering from the Wuhan University of Technology; postdoctorate from Wuhan University and Chinese Academy of Social Sciences. He is also senior visiting fellow at School of Engineering, University of Michigan; professor at Guizhou University of Finance and Economics; founding Director of Research Center for Modern Transport Development in College of Management Cadres of Ministry of Transport and Director of the training office of the Ministry of Transport in support of the western region; chief researcher of Beijing Jiaogan Think tank Information Technology Research Institute; researcher at Research Center for Industrial, Academic Research Issues, Wuhan University; tutor for MPA at University of Science and Technology of China; "Young science and technology talent in traffic" of Ministry of Transport, the excellent manager of science and technology in recognition of the Ministry of Transport and an expert of the famous teacher's group of the Ministry of Transport; a teacher of the training course of the nationwide head of bureau of transportation; core expert of think tank in Guizhou province; core expert of Guangxi Zhuang Autonomous Region; Deputy Director of bureau of transportation in Qinzhou City, Guangxi Zhuang Autonomous

Region and a distinguished expert of the strong division of traffic of Guangxi transportation department. He is in charge of projects such as National Social Science Fund and research projects entrusted by the business department of Ministry of Transport. He is also associated with the Postdoctoral Foundation Project of China, Special Fund Project of China Postdoctoral Foundation, and over 30 National Educational Science Planning Projects. He has published more than 100 academic papers in various key academic journals and more than 20 treatises.

Gusheng Zhu, the professor, editorial Director of the *Journal of Qujing Normal University*; the communication review expert of the National Social Science Planning Office; the leader of "Applied Economics", a key discipline of Yunnan Province; the chief expert of the Yunnan Provincial Key Research Base of Philosophy and Social Sciences; contact expert of Qujing Municipal Committee; major administrative decision-making consultation and argumentation expert of Qujing Municipal Government; master advisor of Shanghai Normal University and Yunnan Agricultural University. He mainly engages in economic and cultural research in the upstream of Pearl River, has participated in the National Social Science Planning Project, the Ministry of Education Humanities and Social Sciences Project research work, has hosted a number of projects, such as the Provincial Philosophy and Social Science Planning Project, the Provincial Social Sciences Key Research Base Project, the University and Province Cooperation Project, the Provincial Education Department Research Fund Project and many other research projects, and has won a series of honorary titles, such as the outstanding professional and technical talents with outstanding contributions at the provincial level, the "Excellent Editor"of journal of local colleges and universities, and won special government allowance for outstanding backbone teachers at municipal level.

Contents

Foreword by Ni Jun v

Foreword by You Zheng ix

Preface xiii

About the Authors xvii

Introduction xxiii

Part I Cognition: Understanding Big Data **1**

Chapter 1 What is Big Data Transportation? 3
1.1 Big Data Transportation and Internet + 3
1.2 Big Data Transportation and Intelligent Transportation 7
1.3 Big Data Transportation and Cloud Computing 11
1.4 Big Data Transportation and Artificial Intelligence 16

Chapter 2 Big Data Transportation: Getting Closer to Life 19
2.1 Self-Service Travel of Big Data Transportation 19
2.2 High-Speed Railway Travel of Big Data Transportation 24
2.3 Highway Travel of Big Data Transportation 28
2.4 Civil Aviation Travel of Big Data Transportation 33
2.5 Freight Transportation of Big Data Transportation 38

Chapter 3 Strategy Blueprint of Big Data Transportation 41
3.1 Big Data Transportation in the United States 42

3.2 Big Data Transportation in Europe 47
3.3 Big Data Transportation in Japan 51
3.4 Big Data Transportation in China 55

Chapter 4 Big Data Transportation: Eight Innovative Modes 61
4.1 Feature 1 of Big Data Transportation: Shared Economy 61
4.2 Feature 2 of Big Data Transportation: Full Utilization of Resources 65
4.3 Feature 3 of Big Data Transportation: Precise Demand 66
4.4 Feature 4 of Big Data Transportation: Real-Time Control 71
4.5 Feature 5 of Big Data Transportation: Efficiency and Convenience 73
4.6 Feature 6 of Big Data: Intelligence 75

Part II Application: Practical Implementation of Big Data Transportation 79

Chapter 5 The Entrance of Big Data Transportation 81
5.1 T-Union 81
5.2 GPS and BeiDou System 85
5.3 Internet of Vehicles 88
5.4 Road Network Monitoring 91
5.5 Electronic Navigation Map 94
5.6 ETC 97

Chapter 6 The Platform of Big Data Transportation 101
6.1 Expressway Management Platform 101
6.2 Accident Detection Platform 105
6.3 Passenger Flow Detection Platform 108
6.4 Traffic Law Enforcement Platform 111
6.5 Logistics Information Platform 115
6.6 Emergency Command Platform 118

Chapter 7 Big Data Transportation and Logistics 123
7.1 Big Data Transportation and Transportation Organization 123
7.2 Big Data Transportation and Rural Logistics 126
7.3 Big Data Transportation and Urban Delivery 128
7.4 Big Data Transportation and Express Logistics 131
7.5 Big Data Transportation and Cold Chain Logistics 134
7.6 Big Data Transportation and Multimodal Transportation 137
7.7 Big Data Transportation and Cross-Border Logistics 140

7.8 Big Data Transportation and Aviation Logistics 143
7.9 Big Data Transportation and Non-Truck Operating Common Carrier 145
7.10 Big Data Transportation and Logistics Costs 148

Chapter 8 Big Data Transportation and Transportation Planning 153
8.1 Big Data Transportation and Route Planning 153
8.2 Big Data Transportation and Network Line Planning 157
8.3 Big Data Transportation and Road Planning 160
8.4 Big Data Transportation and Station Planning 163
8.5 Big Data Transportation and Hub Planning 166
8.6 Big Data Transportation and Capacity Planning 169
8.7 Big Data Transportation and Emergency Planning 172

Chapter 9 Big Data Transportation and Major Activities 177
9.1 Big Data Transportation and Beijing Olympic Games 177
9.2 Big Data Transportation and Shanghai Expo 182
9.3 Big Data Transportation and Hangzhou G20 Summit 186
9.4 Big Data Transportation and One-Belt-One-Road International Forum 191

Chapter 10 Big Data Transportation and Traffic Management 195
10.1 State Diagnosis of Big Data Transportation 195
10.2 Information Analysis of Big Data Transportation 198
10.3 Technological Change of Big Data Transportation 202
10.4 Organization Model of Big Data Transportation 206
10.5 Management Model of Big Data Transportation 210
10.6 Credit Model of Big Data Transportation 214

Chapter 11 Future Development of Big Data Transportation 217
11.1 Personalized Service of Big Data Transportation Industry 217
11.2 Expansion Trend of Big Data Transportation Industry 221
11.3 Investment and Financing of Big Data Transportation Industry 225
11.4 Police Guidance of Big Data Transportation Industry 229

Chapter 12 Unmanned Driving: Transportation Revolution Sweeping Across the Globe 235
12.1 Origin, Development, and Application of Unmanned Driving Technology 235

12.2 Global Distribution: Competing for the High Ground of Unmanned Driving 248
12.3 Technology Routes: Key Technologies for the Unmanned Driving Industry 261
12.4 Intelligent Traffic Management Model Based on the Unmanned Driving 275
12.5 Key Factors in Unmanned Driving from Concept to Practice 285
12.6 Application of the Unmanned Driving Technology in Urban Rail Transit 293

References 305

Index 309

Introduction

This book is designed as a popular science book on traffic big data. It first gives readers a general description of traffic big data, and then introduces the development of traffic big data in various countries from a global perspective, including strategies, status, actions, advantages, etc., and finally explains related concepts and technical base, expounding on the broad prospects for traffic big data innovation. After explaining the concepts, it explores how to make use of traffic big data from the practical point of view. First, it explores the sources of big data, which is the basis, and also analyzes the purposes of the Internet enterprises' access disputes, followed by the application of traffic big data platform. Finally, it elaborates on the big data's pain points, cases of application, business models, and operation experience in the logistics, public transport, and traffic accidents of the traffic industry.

This book is positioned as a popular reading for big data transportation science.

It aims at having an intuitive scene perception of big data transportation. It introduces the development of big data transportation in various countries from a global perspective, including strategies, current situations, actions, advantages, etc., clarifies the relevant concepts and technical basis, and supplements the broad prospect of big data transportation innovation from the perspective of innovation. This helps us get a basic understanding of big data transportation.

Based on the cognition of the big data transportation, we should consider how to apply it from the perspective of practice. The source of big data is the basis of the application of the big data transportation. It also analyzes the purpose of the competition between Internet enterprises for the entrance. After discussion of the data, the application of big data transportation platform will be explored. The specific field of big data which involve the pain points, application cases, business models, operation experience, logistics, public transportation, and transportation accidents in the transportation industry has been described as well. More transportation platforms at different levels can be added, such as county-level, third and fourth-tier city level, so that people at different positions can have their own reference and enlightening cases.

Part I

Cognition: Understanding Big Data

Part I covers the understanding big data transportation that includes various application scenarios of big data transportation, the development status of big data abroad, the clarification of the concept of big data transportation, the security issues, and the essence of innovation.

Chapter 1

What is Big Data Transportation?

The so-called big data transportation means that through the Internet technology and big data technology, the data generated and precipitated in the transportation industry are processed, analyzed, and operated by using data processing tools so as to produce more effective, convenient, and high-value transportation, optimization, and governance programs, bring convenience to people's travel, improve transportation efficiency, save energy resources, reduce pollution emissions, and optimize the industrial structure.

1.1 Big Data Transportation and Internet +

1.1.1 *Big data is coming fast*

As the industrial society moves toward the information society, all human achievements are stored and transmitted in the form of binary information, and the information is converted into digital form. Some scholars have divided the informationization process of human society into three eras, namely the computer era, the Internet era, and the big data era.[1] With the emergence of the information technology revolution, the improvement and popularization of information technologies such as the Internet, mobile Internet, social networking, Internet of Things, Internet of

[1]Xu Jihua, Feng Qina, and Chen Zhenru: *Intelligent Government: The Coming of the Era of Big Data Governance*, CITIC Press, 2014 edn., p. 11.

Vehicles, intelligent phones, and tablets, human-produced data are growing exponentially.

The information chart released by MBA online website shows that every day 294 billion emails are sent out, 2 million blogs are published online, 250 million photos are uploaded on Facebook, 8.64 million hours of videos are uploaded on YouTube, and 187 million hours of concerts are played on Pandora, the streaming music website.[2] According to IBM's analysis, 90% of all the data obtained by the entire civilization of mankind were generated within the past two years. By 2020, the scale of data generated in the world will reach 44 times that of today.

According to a research report by the International Data Corporation (IDC), in the next decade, global big data will increase by 50 times, and by 2020, the world will reach 35 zebits (about 3.5×10 bits) of data information amount. There are many types of data that are emerging. The data are rapidly transmitted in communication networks, and the number of data to be developed is increasing. Human society has officially entered the era of big data.

1.1.2 *Big data transportation is perfect*

With the gradual improvement of big data technology, the integration and use of big data in various industries have become more and more in-depth. Today, from national policies to commercial marketing, big data can be seen everywhere and the value of big data is becoming more and more prominent. Transportation, as a natural field of data precipitation, is closely related to big data. The big data play a crucial role in a mobile phone map or a government traffic management platform.

With the continuous integration, upgrading, and development of the Internet and industry, the transportation industry is increasingly connected with data in various industries, and the multitude of information has brought opportunities and challenges to the development of big data transportation. Seizing the wave of informatization and digitization is the key to realizing the transformation, optimization, and development of the transportation industry.

The big data transportation collects big data sets and processes, analyzes, and operates them with data processing tools. As a natural big data application industry, the transportation industry has a large number of data sources and a

---

[2] *Ibid.*, p. 13.

wide variety of data, and the inconsistent economic development in various regions has caused difficulties in data collection and integration of the industry management system, inefficient or even invalid data processing, and increased costs due to rapid data growth. Challenges and opportunities always follow, but the big data transportation industry conforms to the economic and social development, bringing convenience to people's travel, improving the transportation efficiency, saving the energy resources, reducing the pollution emissions, and optimizing the industrial structure.

The close integration of big data and transportation has a close relationship with the intelligent development of China's automobile industry. On September 9, 2017, the Ministry of Industry and Information Technology reported that the first draft of the *Management Specification for the Adaptability Verification of Intelligent Connected Vehicles on Public Roads (Trial)* has been completed and is being revised. As the world's largest automobile market, China's Internet intelligence is developing rapidly, such as BYD, Cheetah, and Dongfeng Qichen without exception, among which "intelligent interconnection" is the focus. The new onboard interconnection system, language control, and other aspects improve the human–computer interaction; the navigation can be woken up at any time during driving, and the language navigation, phone calls, etc., can be realized, all of which need the support of big data technology. Increasing the content of automobile technology can also effectively monitor the road information at all times, so the combination of big data and transportation management is not only limited to high-end applications in the industry but also connected with ordinary people.

Meanwhile, the increasing number of automobile in China and the rapid construction of highways have also increased the application value of big data. By the end of 2016, the total mileage of highways in Mainland China had been 131,000 kilometers, ranking first in the world. Vehicle travel ranked first in all kinds of transportation modes, with a year-on-year growth of 17.6%. The number of motor vehicles in use exceeded 300 million, and the growth rate in the first half of this year was 3.18%. Therefore, from industry to society, it is more and more necessary to combine big data with transportation management.

1.1.3 *Internet + transportation is at the right time*

On March 5, 2015, Premier Li Keqiang first proposed the "Internet +" Action Plan in the *Government Work Report* (Figure 1.1). The integration and

Figure 1.1 The third session of the 12th National People's Congress opened in the Great Hall of the People in Beijing. Premier Li Keqiang made a report on the work of the government.

Source: The Central People's Government of the People's Republic of China.

development of the Internet and traditional industries will entirely transform the traditional industries and generate new formats. The collision between the Internet and transportation has also formed a new pattern of "rational distribution of online resources and efficient and high-quality operation of offline". The Internet + transportation model has seen rapid development driven by national policies.

Today, the "Internet + transportation" model is gradually improved. For example, the Railway Department has introduced a method for booking train tickets online, allowing consumers to use computers and mobile phones to buy train tickets through the Internet without leaving the house, eliminating the hassle of waiting in long lines in the past. The Civil Aviation Department has realized the online booking for an early time. On the basis of accumulated data, after big data analysis, consumers can use mobile APP to realize functions such as purchasing tickets and check-in through mobile and flight status checking. Vigorously promoting the development of ETC network of expressway is the measure to promote the network in the aspect of highway. In addition, people are more and more inseparable from the navigation system and taxi software.

The development of big data transportation and the Internet + transportation model has gradually become the content that people are accustomed to in their lives.

1.2 Big Data Transportation and Intelligent Transportation

1.2.1 *What is intelligent transportation*

The concept of intelligent transportation was proposed very early. As early as 1990, the Japanese Iguchi Yaichi first proposed the concept of intelligent transportation system (referred to as ITS). But at that time, ITS did not attract widespread attention. It was not until 1994 when the United States changed its transportation research organization intelligent vehicle highway society (IVHS) to ITS America that the concept of ITS was paid attention to.

Over the years, with the development of Internet technology, big data technology, cloud computing, blockchain, and other technologies, the trend of intelligence is more and more obvious, and the proportion of intelligence in the field of transportation continues to rise. The world's important economies, including the United States, Europe, and China, have paid more and more attention to intelligent transportation, and have invested a lot of manpower and material resources for its development.

So far, there is no authoritative definition of intelligent transportation, and various researchers have their own understanding of it. However, we can understand this concept with the help of the definition of intelligent transport given by the President of the first ITS World Congress.

The intelligent transportation system is to effectively integrate and apply advanced information technology, communication technology, control technology, sensor technology and system comprehensive technology to the ground transportation system on the basis of relatively complete road infrastructure so as to establish a real-time, accurate and efficient ground transportation system that plays a role in a large range.

In general, the composition of the intelligent transportation system has seven components, namely, advanced transportation information system (ATIS), advanced transportation management system (ATMS), advanced public transportation system (APTS), advanced vehicle control system (AVCS), freight management system, electronic toll collection system (ETC), and emergency medical system (EMS). Together, these seven components form a three-dimensional intelligent transportation system that affects every aspect of people's lives.

For Xiong'an New District newly approved by the Central government, the government is taking the lead in the construction of intelligent public transportation as the main mode of travel.

According to the WeChat public account "Xiong'an Release" on October 14 in 2017, Xiong'an New District held a docking meeting with Baidu Group. Both parties docked on the potential cooperation fields such as big data, artificial intelligence, and cloud computing.

Chen Gang, member of the Standing Committee and Deputy Governor of Hebei Provincial Committee, Secretary of the Party Working Committee and Director of the Management Committee of Xiong'an New District, said that Xiong'an New District is a city built from scratch, and to draw this blueprint well one needs to achieve breakthroughs in the integration of information technology and smart city development.

In the future, Xiong'an New District will implement a mode of travel based on intelligent public transportation and personalized driving behaviors of unmanned private cars, which will constitute the future road network structure and space allocation model of Xiong'an New District. One of the core concepts of the planning and construction of Xiong'an New District in the future is to return the city to its people, from changing the traditional way of designing the city and road according to the scale of cars to designing according to the scale of people, thereby returning the city to its people, and returning the road and space to the people. Baidu group has leading advantages in smart city construction, big data, automatic driving, etc. We welcome Baidu to prepare to build an open and international national-level artificial intelligence laboratory and high-end application research institution in Xiong'an New District, especially to build a demonstration area of driverless high-tech industry.

Lu Qi, President and Chief Operating Officer of Baidu Group, said that Baidu Group hopes to deepen its cooperation with Xiong'an New District in areas such as smart city planning and design, information infrastructure construction, and smart industry pilot demonstration.[3]

From the construction of Xiong'an New District, we can see that in the traditional transportation system, people, vehicles, and roads cannot be unified, coordinated, and integrated due to technology and other factors, and the efficiency and safety of the entire transportation system cannot be guaranteed. In the intelligent transportation system, we use information technology, data communication transmission technology, electronic sensor technology, satellite navigation and positioning technology, electronic control technology, computer processing technology, and

[3]From the WeChat Public Account, "Xiong'an Release".

transportation engineering technology, and these technologies can achieve integration and fusion. The shortcomings and difficulties of the traditional transport system have been overcome one by one, and people, vehicle, and road have become more closely coordinated. The transportation efficiency has been improved, the transportation safety has been ensured, the transport environment has been greatly improved, and the energy efficiency has been greatly enhanced.

1.2.2 *Big data in intelligent transportation*

Intelligent transportation is different from traditional transportation because of its intelligence. Advanced information technology, data communication and transmission technology, electronic sensor technology, control technology, and computer technology, as well as big data technology ensure that the transportation system can play a more powerful role in the command and deployment, can more proactively predict risks and prevent various emergencies than the traditional transportation system does.

Big data technology is providing strong support for this value.

With the rapid development of social economy, the transportation around the world will face more and more complex situation. With the increase of motor vehicles, the development of urbanization leads to the imbalance of traffic road system, and the traditional transportation information system cannot meet the current traffic situation. For example, in cities like Beijing, Shanghai, Shenzhen, and other first-tier cities, traffic congestion has become a problem for the whole city governance.

However, by aggregating all kinds of scattered transportation data to form a kind of data pond and processing data with big data technology, we can analyze some transportation trends, demands, difficulties, etc. Then through the refined data processing and image processing, transportation suggestions and decisions are formed. For instance, some people have a specific information demand for a certain traffic section. Although the demand seems to be small in number and share, a specific and personalized service can be formed through the processing of big data technology and aggregation of other similar demands. Based on this, the transportation management system will be more intelligent.

The big data make transportation management more intelligent, which is reflected in several aspects.

First, the data across administrative regions will be integrated and the transportation management will break through the limitations of previous administrative divisions and establish an overall management awareness of chess. This will help the government departments to carry out the construction of transportation facilities more effectively and the transportation route planning more reasonably.

Second, big data technology can reasonably allocate the transportation information resources and improve the efficiency of information utilization. At present, the responsibilities of China's transportation departments are overlapping, some of which are not clearly defined. By using big data technology, we can reasonably allocate the human and material resources, assist relevant departments to make a more reasonable overall plan, so that the utilization rate of transportation information is higher and the transportation management is more intelligent.

Third, big data technology has changed the thinking mode of transportation management in the past. In the traditional thinking for transportation management, when encountering traffic jam, the measures of widening the road and increasing the mileage are taken to improve the traffic capacity. In fact, such management is often short-sighted. Land resources are non-renewable resources, and widening roads can only waste more land. Moreover, in the peak period of traffic, widening roads cannot solve all problems. Through the analysis and integration of big data technology, we can have a clear understanding of the time period, location, and cause of traffic congestion, so as to adjust measures to local conditions, conduct reasonable dredging and governance to reduce the dependence on land, space, and other resources, and achieve a virtuous cycle of transportation facilities and systems.

We can understand the value of big data in intelligent transportation by observing the construction of intelligent transportation in Changsha.

At present, Changsha mainly constructs two platforms, six systems and five application software in transportation informatization. In March 2016, the third phase construction of intelligent transportation was started, which was undertaken by Qingdao Hisense Network Technology Co., Ltd. and had data exchange agreements with Baidu, Amap, and other Internet companies, so as to use scientific and technological means to boost the transportation management.

Currently, all traffic police businesses in Changsha can be managed through mobile phone APP, including car price, penalty for driving offense points and query, etc. In the next step, mobile phone APP can be used to query the surrounding parking lot information and the capacity for parking.

The two platforms of intelligent transportation management are command and dispatching platform and video network sharing platform. The six systems are high-definition electronic police system, high-definition television monitoring system, intelligent traffic signal control system, traffic flow collection system, network security and storage system, and secondary and tertiary sub-control center transformation. The five application software systems are road traffic PBE collected, processed, and analyzed in real time; the value of intelligent transportation will no longer exist.

Third, the data can be widely shared. The information resources of the transportation industry are complex and valuable. Not only the government departments have a great demand for them but also the related enterprises have a strong need. The collection of these data and information needs to be completed by government departments and enterprises together, and the purpose of these also needs to be shared. Otherwise, a single island of information will be generated, and the unified allocation and use of information resources will not be realized.

Fourth, the intelligent transportation must be highly stable. Especially for traffic command and guidance, if high stability will not be achieved, high safety, high efficiency, and high accuracy of transportation cannot be guaranteed. Urban traffic congestion may occur, resulting in waste of resources and environmental pollution.

1.3 Big Data Transportation and Cloud Computing

As we all know, big data has integration advantages and combination efficiency at the application level. With the support of big data technology, information resources in any field can change the situation of information decentralization, single information content, disorder and isolation.

However, in order to store, analyze, and share massive data, we must have a powerful platform provided by cloud computing technology. Cloud computing technology can calculate a large number of data quickly, and users can use the data conveniently. It can even let users experience 10 trillion times of computing power per second. With such powerful computing power, cloud computing can simulate nuclear explosions and predict climate changes and market trends. Users accessing the cloud computing platform can calculate according to their own needs through computers, laptops, mobile phones, and other ways.

For the transportation field, big data and cloud computing technology can help establish a comprehensive three-dimensional transportation information system. Any node in the transportation system can efficiently calculate the information and call the data. The whole transportation participants will be incorporated into the system, playing out the overall function of transportation. For example, for infrastructure construction, traffic information release, value-added services for transportation enterprises and traffic command, the big data and cloud computing can provide decision-making and simulation, which is very efficient.

Besides, cloud computing technology has the characteristics of generality, high scalability, on-demand services, extremely cheap, and it is the inevitable demand for the development of big data transportation. The development speed and quality will be faster and higher in the big data transportation in the construction process, with the support of cloud computing.

Guizhou Province, as a big data construction base in China, is also at the forefront of exploration in big data transportation and cloud computing. From this, we can also see the great value of cloud computing in the development of big data transportation.

... I would like to take this precious opportunity to share with you the practice and exploration of Guizhou traffic police in the field of road traffic safety by using big data and cloud computing technology.

At this time last year, Premier Li Keqiang proposed at the executive meeting of the State Council to support the research and development of key technologies and major project construction of cloud computing. Today, Guizhou traffic police has built the largest private cloud platform for government affairs in China. This time last year, the State Council was using big data to develop market-oriented and individual credit investigation business and accelerate the system construction of credit evaluation system business. This time last year, the major traffic accidents on Guizhou highway were frequent. This year, the number of accidents on the province's 5,000-kilometer highway decreased by 11, and the number of death decreased by 47 on a year-on-year basis, setting a record for the best accident prevention since the opening of the highway in Guizhou.

This time last year, Guizhou's rural road traffic safety supervision work was also plagued by the problems of having a mechanism without implementation, institutions without personnel, measures without means, or responsibility without accountability. This year, Guizhou traffic police has built a data cage and safety supervision cloud platform for strengthening rural road traffic safety at the provincial, municipal, county, township, village, and group levels.

All the changes come from the construction of Guizhou traffic police cloud and the support for intelligent transportation from cloud to end after the completion of cloud computing platform. By relying on big data and cloud computing technology, with the modern concept of governance and modern information technology, the Guizhou Internet + road traffic management model is gradually formed. At a time when big data, cloud computing, and Internet+ have become Internet buzzwords, Guizhou traffic police have come out of the confusion of how to train cloud platforms.

Our experience is that we are not superstitious about cloud services because cloud services are not all-powerful. Without standards building for business applications, without meticulous data framework design, simply aggregating cluttered and chaotic data and putting it on a cloud platform can also create dangerous data services.

Our specific approach is based on the guidance of the Ministry of Public Security and the Public Security Cloud Computing Construction Center. The functional requirements of the public security cloud platform are guided by the problems existing in the informatization construction of Guizhou public security, with the main business of Guizhou traffic police as the focus. It has established strategic cooperation with the Traffic Management Research Institute of the Ministry of Public Security, absorbed mature cloud computing technology from Alibaba, adopted Huawei hardware, rented the computer room and equipment of China Telecom, and built 618 servers, 46 network switches, and 12 PB storage capacity according to the sequence of business application + data framework. This cloud platform is compatible with various traditional data types, uses inexpensive X6 servers as video and picture storage units, and can store 75 HD pictures, with 10,000 cores of computing power, 20,000 pictures per second and processing, distributed database, and big data online and offline computing all kinds of cloud services, which cracks the four major problems that constrain the construction of the

Guizhou traffic police information technology and that cannot be stored, run, well used or shared.

After building the cloud computing platform, our practical application can be summarized in two sentences. The first is to realize cross-border integration and data sharing with big data resource pool. Cloud platform construction enables us to realize big data resource pooling, provides data basis for information resource service, and provides solid service guarantee for the information resource sharing and big data application of cross-regional, cross-industry, cross-department, and cross-police type. The internal sharing of public security information resources lays a solid foundation for the construction of three-dimensional public security prevention and control system. Information resources are shared externally with all sectors of the society. Data from various industries and departments are deconstructed, intersected, and integrated in the big data resource pool. Through convergence, cleaning, and sorting, a standardized data resource directory is formed and aggregated into sea data, and a good big data ecosystem is gradually formed.

The second is to realize responsibility sharing and security co-governance with the support of cloud platform. Aiming at the management difficulties of key vehicles, expressways, and rural roads, we have developed three regulatory cloud platforms supported by cloud computing and cloud platform, solving the management problems. In response to the problem of the lack of key management means to the key vehicles that transport dangerous goods, we have built a Guizhou road traffic safety comprehensive supervision cloud platform, relying on cloud computing technologies such as various financial relational databases to achieve accurate early warning and dynamic intervention of vehicle dynamic violations and monitor every move.

Since the platform was launched, the provincial regulatory authorities at all levels have logged in and used the platform 189,000 times, and the enterprise users have logged in and used the platform 570,000 times. The traffic police, transportation, safety supervision, and enterprise have realized the real-time sharing of information, solved the problem of information asymmetry in safety supervision, and clarified the responsibility of safety supervision of all departments and the management responsibility of enterprises.

In combination with the key driver credit system, we have checked more than 6 million drivers in the province, realizing the precise control of key drivers and playing an unprecedented management effect. Now what the big screen shows is the relevant data such as the dismissal of key drivers by enterprises in the whole province this year and the lifetime prohibition of key drivers by traffic police, aiming at the problem that the expressway cannot achieve the management effect relying on the traditional police operation mechanism.

We have set up an intelligent comprehensive supervision cloud platform, relying on the operation of the associated database, using video image big data analysis technology to control and collect key vehicles, implementing accurate early warning for key violations such as overspeed at two stations, implementing four pushes of illegal data information, dynamic intervention to eliminate in five steps, and accurate control of key vehicles. In the first three quarters of this year, more than 340,000 illegal vehicle information were pushed, and five steps were taken, including front-end collection of road network discovery, dynamic tracking, safety guidance and site punishment. More than 374,000 traffic violations were investigated and dealt with on the spot by law enforcement station and service areas.

In view of the problem of long occupation of rural roads and serious shortage of management power, we have independently developed a cloud platform for comprehensive supervision of rural road traffic safety, extending the tentacles to rural roads accounting for 95.5% of the total mileage of the province. The platform is used to collect road traffic safety data and key period information of rural inhabitants, vehicles, roads, climate, etc. The big data mass analysis technology is used to give an early warning on the abnormal situation of drivers' vehicles in each township, track the investigation of hidden danger road sections, and automatically push the early-warning information regarding the weddings or funerals, folk activities, national festivals, and so on to the five-level responsible person, implementing the responsibility of supervision to the individual. Through the comment and command, pre-warning, in-process supervision, and post accountability, the responsibility of territorial management of Party and government at all levels has been truly implemented, and the goal of responsibility sharing and safety sharing has been achieved.

Guizhou traffic police cloud which was built and applied in the last year has made little achievement, but has found the right way, a good start, and a good step forward. With the rapid development of big data, cloud computing, Internet of Things, and other new technologies, the vigorous development of intelligent transportation has brought a huge impact on the theory and practice of traditional transportation. In the next step, Guizhou traffic police will deeply study the advanced experience of public security road traffic management in other provinces, absorb the research results of intelligent transportation from this forum, continue to expand the application of big data under the leadership of the Traffic Management Bureau of the Ministry of Public Security and the strong support of Wuxi Research Institute, explore the precise governance and personalized management scheme of road traffic safety supervision with cloud computing technology, strive to achieve accurate supervision and targeted treatment of the problems of road traffic safety supervision, and further promote a number of transformative achievements to improve traffic safety and traffic management, creating safe traffic, harmonious traffic and convenient traffic, and making greater contributions to serve the road traffic management, public security and economic and social development.[4]

1.4 Big Data Transportation and Artificial Intelligence

Artificial intelligence has become very popular in recent years. Regardless of the automatic driving technology or the electronic traffic police, the artificial intelligence is indispensable. Generally speaking, artificial intelligence simulates, extends, and expands human intelligence by means of Internet, big data, cloud computing, robotics, and other technologies and means. With the help of artificial intelligence technology, human beings hope that machines can do the work that only the human beings can do. In fact, artificial intelligence technology has developed rapidly in recent years. In some fields, artificial intelligence has surpassed some human capabilities, such as alphago's victory over human chess player Ke Jie.

The application of artificial intelligence in the field of transportation is more extensive and in-depth.

[4]The content is quoted from the theme report "The Practice and Application of Big Data and Cloud Computing Technology in Guizhou Public Security Traffic Management Work" by Wang Jianwu, Director of Traffic Management Bureau of Guizhou Public Security Department, at the "10th China Intelligent Transportation System Market Seminar in 2015", which has been deleted and modified.

At the 6th China Intelligent Transportation System Market Seminar that ended on March 23, 2017, several industry experts said that the construction of domestic intelligent cities still remained in the traditional vertical application and lacked in-depth application. Through artificial intelligence + big data + transportation, the urban intelligent transportation system has "external brain" for the construction and development of intelligent city through self-regulation.

"Big city problems" such as traffic jams have been criticized for a long time. How to solve the problem of traffic jams has become an important issue for urban management decision makers. Some people in the industry think that how to improve the efficiency at intersections through signal control is the key to solving the problem, which is also the most basic key link in the construction and development of intelligent cities.

Recently, based on the data mining and in-depth learning of artificial intelligence in the construction of intelligent city, Hisense has demonstrated to the outside world its early layout breakthrough application scheme in the field of transportation. It is said that the "data cube" developed by Hisense can complete the visual analysis of 1 billion large-scale transportation data in 30 seconds, and realize the transportation prediction based on deep learning.

"Based on massive transportation data, we analyze the regularity and similarity of transportation operation, build an intelligent learning model, and predict traffic flow, congestion and other traffic parameters through machine learning". Ma Xiaolong, an expert in the information control for urban transportation from Hisense, said that through real-time monitoring and analysis of road traffic flow, according to dynamic traffic data, automatic switching and allocation of signal time, the most intuitive change is that the time of traffic lights is no longer fixed, or even people do not stop during green lights in the whole process.

It is understood that the more desirable state of transportation described earlier has been achieved. In the adaptive signal control system of Hisense, according to the recommended speed, "all the way green light" can be achieved. Data show that in Qingdao, Shandong province, the completion of the construction of the intelligent transportation system has reduced the peak duration by 1.48 hours, increased the average speed ratio by 9.71%, saving the transportation time for vehicles by 107,938,000 hours.[5]

[5]The content is quoted from "Artificial Intelligence + Transportation Big Data, Intelligent City Construction Has 'External Brain'" by guangming.com.

The aforementioned application is only a microcosm of the application of artificial intelligence transportation in China. All major cities in China are constantly exploring and applying artificial intelligence to improve transportation efficiency to make urban traffic smoother. In addition, China has issued many policies and regulations to encourage the development of artificial intelligence and big data transportation. In the future, China's big data transportation and artificial intelligence transportation will have a new look.

Of course, from a global perspective, the important economies in the world are developing artificial intelligence. The United States has turned the idea of "smart earth" into a national strategy. Japan and South Korea are stepping into various fields of national life and production through artificial intelligence and big data technology, constantly improving people's living standards. The EU is also constantly focusing on the construction of intelligent cities in the fields of artificial intelligence, such as environment, transportation, medical treatment, and intelligent building, which are all developed earlier and deeper than China.

Although artificial intelligence is at the development stage, its future prospects are very bright. The transportation field is generating a series of data every moment, and this huge amount of data, through the analysis and learning of artificial intelligence, will be better for the transportation participants to draw a picture. For the government transportation sectors, railways, airlines and tourism agencies, the precise customer base can be found to provide a more individualized service.

Chapter 2

Big Data Transportation: Getting Closer to Life

Many people feel that big data transportation is far away from their lives, however fact is, it is all around us. The daily travel information forecast, shared cycling, and subway bus scheduling are closely related to the big data transportation. It has a profound impact on our lives and has greatly changed our travel behavior and habits.

2.1 Self-Service Travel of Big Data Transportation

With the increasing popularity of automobiles and the diversification of transportation modes, as well as the development of intelligent transportation, people's travel methods are increasingly personalized. Whether it is transportation to work or holiday trips, self-service travel is becoming more and more popular with consumers. People's sense of autonomy and self-service are getting increasingly stronger, and they no longer need staff to give full guidance during the trip, and they can complete the whole process of inquiry by using their mobile APP.

Looking at the current market, the way people travel on their own involves various means of transportation, including bicycles, cars, trains, high-speed trains, airplanes, etc.

2.1.1 *Self-service travel by bicycle*

Self-service bicycles, also known as public bicycles, are originally bicycle rental services provided by government departments to improve urban transportation, reduce urban congestion, and traffic pollution. At public bicycle rental sites set up by the government, people can open and use bicycles by swiping cards or unlocking by their mobile phones. However, due to the inconvenient deposit payment, the need to return bicycles to the specific places, and the old style of bicycles, the users' experience of the government-led bicycle rental is not good.

In recent years, with the rise of shared bicycles, the self-service bicycle travel has become more and more common. From users' registration, deposit payment, bicycle use, bicycle return, etc., all are completed by the users themselves. The shared bicycle operation team is only responsible for the processing of transportation information data, bicycle quantity scheduling, and bicycle maintenance.

For example, Mobike, OFO, and other shared bicycles that are familiar to everyone are the best ways for people to travel by themselves. Users can get and return the bicycles anytime and anywhere. In short-distance travel scenarios such as subway stations to the company and bus stations to home, the value of shared bicycles is great. Especially in the first-tier cities with dense bus lines and heavy traffic, the shared bicycles play an important role in short-distance travel and provide a faster, more economical, and direct way of travel. Although the shared bicycles, to some extent, form a competitive relationship against short-distance transportation of buses and taxis, it alleviates traffic congestion to a certain extent and reduces environmental pollution, which can be used as a supplement to the "last kilometer" of urban public transportation connection.

In addition, according to the latest "Big Data Observation for 2017 China's Sharing Travel Industry" released by CBNData, in terms of self-service travel, the shared bicycles and other travel modes are subverting and innovating the transportation mode. With the rapid development of shared bicycle industry, the shared bicycle has become the most widely and frequently used mode in China. By 2018, the scale of shared bicycle market in China will reach 10 billion.

CBNData, based on the data analysis of OFO shared bicycle platform, says that the shared bicycle solves nearly 80% of users' demand of "difficult travel within two kilometers". It plays the role of short-distance travel and short-distance connection, greatly improves the travel efficiency, effectively supplements the medium and long-distance travel and it is the best choice for consumers for short-distance travel and short-distance connection.

Based on the data analysis of OFO platform, CBNData claims that the shared bicycle is the best choice for consumers to travel in short distance and connect in short distance. For example 36.2% of OFO users ride within 500–1,000 meters and 26.4% of OFO users ride in the range of 1–2 kilometers.

In Shanghai, nearly 20% of OFO users travel within one kilometer of the subway station every day. The shared bicycles solve nearly 80% of users' demand for "difficult travel within two kilometers", play the role of short-distance travel and short-distance connection, and effectively supplement the medium and long-distance travel.[1]

2.1.2 *Other self-service travel*

In other aspects of self-service travel, high-speed rail and airplanes have realized self-service travel to some extent, which provides consumers with many conveniences, alleviates urban transportation, and optimizes transportation conditions.

For example, in terms of designated driving, the traditional mode is that the customer places an order, and the platform sends the order to the driver through information matching. In this process, the customer does not have too many independent options, and most of the processes are completed by the platform.

But now the designated driving model has changed, and the right to self-service and autonomy has turned to customers. Customers can use the map software in their smartphones to find the designated driving service. Through the location-defined service, they can choose the designated driver closest to them, choose a driver with a good reputation, and filter drivers according to their other needs. Such a process is done by the customers themselves. The map software provides corresponding information services, but the autonomy is in the hands of the customers.

According to the relevant persons of Amap, although the designated driving service seems simple, it involves the personal and property safety

---

[1]The content is quoted from *China Sharing Travel Report: The Best Choice of Shared Bicycle for Short Distance Travel*, xinhua.com.

of the owner and other vehicles, so the qualification review of the designated driver is extremely strict. The designated driver must have a good character and skilled driving skills. If customers want to travel by themselves, they need to be guaranteed in terms of security.

In the APP of Amap, there is an option to find a designated driver in the "nearby" feature. Amap uses the map and mobile phone positioning to accurately recommend the appropriate five designated drivers to the user. The distance from each driver to the customer will be shown so that the customer can be aware of when choosing, and will not wait too long. Moreover, the information of these designated drivers includes photos, driving age, place of birth, number of designated driving, rating stars, etc. Customers can know their relevant information when they choose by themselves so that they can better arrange their travel.

For another example, in the field of public transportation, especially subway, the degree of self-service travel is getting higher and higher, in particular with the popularization of smart phones, mobile phone self-service recharge, card swiping, and riding are completely self-service process. The promotion of Bluetooth-entering the metro gates for mobile phones in the Nanjing Metro is an excellent evidence for this.

Highlight 1: Entering the metro by mobile phone is fast and convenient, and we may say goodbye to "queuing" — Once the mobile phone is swiped, passengers can enter and exit the station. By turning on "Bluetooth" to turn on the function of "passing through the gate" and interacting with the "gate" of the metro, passengers can get in and out of the gate without buying tickets and avoiding the embarrassment of forgetting to take citizen cards and taking the bus without cash. Now, there is no need to queue up to buy tickets, just use the Bluetooth of mobile phone to enter the gate, which is very convenient.

Highlight 2: Recharge your account with one click and charge anytime and anywhere — The "recharge" function of Nanning Rail Transit APP can help passengers to recharge anytime and anywhere, with the option of setting the amount or inputting it by themselves (the range is 2–500 yuan). The "recharge" function is connected to three online charging methods: UnionPay card, Alipay, and WeChat.

Entering the "bill" function on the APP, users can also query the account balance, details of each consumption, recharge, and consumption statistics within half a year and other information in real time. How much money passengers have left to take the metro can be checked by their mobile phone.

Highlight 3: Metro information can be checked and "housekeeper" can protect the travel — Nanning Rail Transit APP saves the links of ticket purchase and recharge, and passengers can enjoy the high-quality passenger service at any time. The "timetable" can query the round trip time and the first and last departure time of the trains. The "ticket price inquiry" can offer the route and understand the ticket price according to the starting point and terminal point. The "station information" can query the station information, entrance, and exit and the surrounding information of the station. Nanning Rail Transit APP is the attentive travel housekeeper around the passengers.

Highlight 4: The arrival information is known early, and the appointment would not be late — Passengers can view the "arrival reminder" function by swiping the "small clock" to the left on the APP homepage, which is used to locate nearby stations and view the arrival time of the nearest train. Passengers can use this function to locate the distance between themselves and the metro station more accurately during daily travel. Even if they are not in the metro station, they can check the arrival time of the two nearest trains, which can help passengers adjust their itinerary according to the actual situation. The appointment will certainly not be late.

Highlight 5: Multiple features come together, and no one knows you better than the APP — The "discovery" feature can also automatically recommend popular merchants near the metro stations for you, and restaurants and activities surrounding the metro are also listed. In the "nearby merchants", you can choose to switch to other station for query. The travel life of passengers is enriched to drive the business circle around the metro, and "discovering" the smart life on the mobile phone is more wonderful.[2]

With such a convenient public transportation, travelers no longer need to queue up to buy tickets or be anxious when forgetting to take bus cards. With mobile phones, all travel behaviors can be completed, and the travel

[2]The content is quoted from WeChat Public Account "Nanning Railway Transit" (NanningRT): "Farewell To the Queue: Nanning Metro Opens the Era of 'Mobile Phone Bluetooth' Entering the Metro Gates", the content has been deleted and modified.

Figure 2.1 A passenger using Bluetooth sensor to pass through the Metro gate machine.

Source: *China Daily News*.

trajectory can be easily found through mobile phones, which is a win–win situation for metro operators and travelers.

For metro operating companies, a large amount of traveler's behavior data can help them better plan the metro's departure frequency and running time, and can provide more self-services to reduce operating costs. For travelers, any information, track and transportation time can be found on mobile phones. It is very convenient and quick (see Figure 2.1).

The self-service travel is distributed in all links of the transportation system. From short-distance travel to long-distance travel, the value of big data in transportation is becoming more and more important, especially the map software plays a core role in the process of self-service travel. If there is no positioning data, driving, street and other data, the convenience of consumers' life is impossible.

2.2 High-Speed Railway Travel of Big Data Transportation

With the dense distribution of China's high-speed rail network, Chinese people are choosing high-speed rail to travel more frequently. According to the relevant data statistics of China Railway Corporation, since the operation of Beijing–Tianjin Intercity Railway in 2008, by September 30, 2017, China's high-speed rail units have delivered more than 7 billion passengers, with an average annual growth rate of more than 35%. It is important to know that the total population of the world, according to incomplete statistics, does not exceed 7.3 billion people. If we calculate the number of passengers commuting China's high-speed rail at 2 billion per year, by 2018, the total number of passengers delivered will definitely exceed the total number of people in the world.

The speed of China's high-speed railway construction is rapid. A mere five years from 2012 to 2017, a number of key high-speed railway projects such as Beijing–Guangzhou Railway, Shanghai–Kunming Railway, Harbin–Dalian Railway, Guiyang–Guangzhou Railway, Lanzhou–Wulumuqi Railway, and Hainan Roundabout Railway have been completed and opened to traffic, and the "four vertical and four horizontal" high-speed railway network is basically formed. From China's small cities in the north, to the southern seaside of Sanya, from Xinjiang in the west to the eastern coastal areas, the high-speed railway is like a human artery connecting the country. According to the statistics, China's high-speed railways, together with other railways, have formed a rapid passenger transport network of more than 40,000 kilometers, basically covering China's provincial capitals and cities with a population of over 500,000. The Yangtze River Delta, the Pearl River Delta, the Bohai Rim, and other city groups are more closely connected by high-speed railways, and the four major plates in the east, central, west, and northeast have achieved high-speed railway connectivity.

The high-speed railway is so convenient and efficient, people's travel is naturally more intelligent. But the vast network of high-speed railways requires big data technology to unify and control them. Once something goes wrong at one point, it can affect subsequent running of the trains. In previous years, there were frequent instances of high-speed railway scheduling problems causing subsequent trains to be late, but now such news is rare. Because China's transportation network system is increasingly getting better, big data technology is getting more and more powerful, which can guarantee an efficient and complex high-speed railway network.

We can see some clues from the "2018 Spring Festival Transportation Big Data Report" released by Tongtong Travel and China Communications News.

For the 2018 Spring Festival transportation, railways were expected to transport 393 million passengers, up 10% from last year and well above highway transportation by 1.1%. With the shift in people's consumption perceptions and the increase in civil aviation capacity, more and more consumers choose civil aviation during the Spring Festival transportation, which is expected to transport 67 million passengers in 2018, an increase of 14% over the same period last year, has the largest increase among the three major transport sectors.[3]

The number of passengers transported by the railway sector increased by 10% year-on-year, which is inextricably linked to the gradual use of high-speed rail lines. At present, China has more than 2,700 D-series high-speed trains, operating more than 4,500 trains per day, ranking first in the world, and transporting more than 6 billion passengers in total. The Beijing–Shanghai high-speed railway, for example, has safely transported 672 million passengers in its six years of operation.

High-speed railway travel is an important mode of transportation, and its application and practice in big data are reflected in these aspects.

2.2.1 *Intelligence*

Regardless of features such as buying a high-speed railway ticket or checking in when riding the high-speed railway, the high-speed railway network is becoming more and more intelligent. For example, Beijing West Station has opened a human face recognition system; passengers enter the station through swiping ID card, and then enter the gate by the human face recognition system. The ticket check-in can be completed in just 1–2 seconds. These tasks, which used to be carried out by manpower, now have made travel much more efficient.

2.2.2 *Self-service*

Behind the self-service is actually the embodiment of intelligence. The passengers can make inquiries and book tickets through their own APP according to the information on where they want to go, where they want

---

[3]The content is quoted from "2018 Spring Festival Transportation Big Data Report: More Rational Travel of People in Spring Festival Transportation, Huge Potential Demand for Intelligent Transportation Services", china.com.

to transfer trains and when they want to depart, which not only improves efficiency but also saves manpower and resources as well. The high-speed railway has also opened the functions of food ordering and ticket booking, which will make the travelers in the process of travel more convenient and green.

2.2.3 *Low cost*

Although the ticket of the high-speed railway is much higher than the ordinary train ticket, from the overall economic value of society and the value of time, the high-speed railway saves the travel time, improves the economic value of the unit time, and reduces the cost of travel for consumers, as described in the following news.

The development of high-speed railway has also livened up the economy and commerce in the areas surrounding the big cities. Mr. Chen, who works in the financial industry, is processing emails in the waiting lobby with two cell phones and a laptop. Because of his job, he has to commute between Beijing and Wuqing of Tianjin many times a week. Beijing and Wuqing are about 90 kilometers apart, and it takes more than an hour to drive without traffic jams, while the high-speed railway takes only 24 minutes. "Transportation saves both time and labor. Driving a car is quite troublesome, especially in Beijing. When you drive to Beijing, you may have traffic jams. There's no risk of traffic jams by taking railway, so you don't have to delay your work. If we drive a car, the cost will be too high. In fact, the fuel fee and the bridge fee are far beyond the high-speed railway ticket. I only spend more than 70 yuan back and forth. If we drive, the cost must be higher than that".[4]

2.2.4 *Safety*

In terms of punctuality and personal safety of passengers, high-speed railway travel has more obvious advantages than air and highway travel.

China has the world's longest total length of high-speed railway and the highest frequency of travel for the public. High-speed railway travel

---

[4]The content is quoted from Teng Yun, Su Jiao, and Chen Yan: "Chinese People Summarize Three Characteristics of High-Speed Railway: Convenience, Efficiency and Safety", peoplerail.com.

Figure 2.2 High-speed rail station.

Source: *The People's Daily Network.*

will continue to change the way people travel, further improve the efficiency of transportation, facilitate the life, and reduce environmental pollution (see Figure 2.2). All this is closely related to the transportation big data network. It can be said that the high-speed railway network is a data network. Through the support of massive data, high-speed railway can be extremely convenient, efficient, and safe.

2.3 Highway Travel of Big Data Transportation

The highway travel is one of the most important ways for people to travel in China. The highway travel concentrates on short distance travel. People prefer to highway travel when the travel distance is less than 200 kilometers. According to the "2018 Spring Festival Transportation Big Data Report" released by Tongcheng Travel and China Communications News, the travel distance within 200 kilometers and above 800 kilometers is the highest proportion, which is 33.5% and 30.1%, respectively, meaning that the role and value of the highway travel is great.

In our everyday life, if we look from the means of travel, there are generally several kinds of highway travel such as the car, bus, motorcycle, etc. In terms of the way of travel, there is a wealth of ways to get around, such as self-driving, carpooling, renting a car, taking a bus, taking a taxi, taking Didi taxi, etc. From the travel scene, it includes urban road travel, highway travel, national highway, provincial highway, county highway travel, etc.

No matter what kind of highway travel it is, it can be said that it is the top priority in transportation management. In the transportation management, the highway travel data are also the most complex and scattered, but it has the most value of analysis and research.

2.3.1 *Highway travel under the big data*

In recent years, the mileage of highway construction in China has been increasing, and the challenges of road network maintenance and management have become increasingly prominent. The work of toll collection, monitoring, communication, and other fields on the highway is not intelligent enough. A large number of human and material resources need to be put into it, which increases the cost of highway management.

At present, the road condition information of many high-speed sections still relies heavily on manual reporting or display screen. As a result, road information cannot be fed back in real time, and the blindness and uncertainty of people's travel will increase. Especially in case of severe weather such as heavy fog, ice, and snow, if the highway management department cannot release the information on time, and cannot monitor and command the traffic conditions of each road section through data, it may cause major safety risks.

We can simply divide big data management of highways into safety management, road administration management, and service management.

In terms of safety management, the highway has a large amount of decentralized traffic data, advanced electronic sensor technology, video surveillance equipment, near-field communication equipment, etc. These devices and data, if utilized wisely, can help management department build a large scale, all-encompassing, real-time, and an efficient safety management system. Through big data analysis and program development, this system can monitor the traffic flow, weather conditions, road conditions, and vehicle accidents and other information, and can send the information of certain road sections prone to accidents to the public through the Internet of Vehicles, traffic radio, WeChat, and other channels.

Of course, map software such as Baidu Map and Amap are now connected to government departments' advanced transportation information systems (ATIS), which can also be pushed in real time through map software. In 2017, Beijing's highway road administration system first tried highway robotic patrols. Such attempts will be a big trend in the future.

In special areas such as tunnels, for example, the lights in the tunnels are adjusted at the right time using big data via a dedicated traffic zone controller. If the sun is shining brightly outside, the lights will be brighter in the area just inside the tunnel entrance, and then gradually return to normal, allowing the driver to adapt naturally to the tunnel environment. When people encounter an accident on the highway, people can ask for help through a visual telephone at the roadside, and can have a video conversation with a professional, which greatly improves the efficiency of rescue.

In terms of road administration management, with the help of big data, the highway road administration management department can reduce the investment in human and material resources and improve management efficiency. For example, the management department integrates all waveform boards, 100-meter piles, kilometer piles, electronic information boards, cameras, etc., on both sides of the highway and the central isolation zone into a unified management system, so that if any related equipment is damaged or lost, the system can be accurately located, replaced, and updated quickly. For highway buildings, overpasses and underpasses, it is also possible to manage and monitor them in the form of data to maintain and protect highway assets more rationally.

The value of big data is even more evident in service management, such as highway tolling and service area management. For tollbooths, video and documentation are left behind as each vehicle passes, and these raw resources are extremely important data resources. On the basis of these data, data processing, analysis, and mining can form a series of effective and high-value data assets. For example, if people encounter a traffic accident and escape, the big data materials of the tollbooth is very valuable, which can help the police to investigate and locate the relevant people. In case of overspeed, the management system can actively intercept the relevant vehicles at the tollbooth with the help of big data to ensure traffic safety.

For the service area, the data such as the length of time and the number of vehicle stops are of great value to management for safety supervision. It is also an invaluable resource for business decisions related to service enterprises in the service area. Which service area has a large demand for a certain service in which time period and which area has a preference for a certain commodity or service can be obtained through big data analysis. On the one hand, this will reduce the waste of inventory in service companies, which can better serve travelers during special times such as holidays, and on the other hand, it will provide personalized services to travelers and improve their convenience.

In the future, with the popularization of new energy vehicles, the service of charging equipment such as charging piles will be an important service content. According to the big data analysis, the service area can arrange the charging equipment and service facilities of electric vehicles in advance.

2.3.2 *Urban road travel under big data*

The urban travel is more complex than the highway travel, and it requires more transportation management systems to match. In addition to the control system of the vehicle itself and the Internet of Vehicles, the management-led systems are important players in urban transportation.

(1) **Transportation information system.** This system is based on a well-developed information network, where traffic participants provide real-time local traffic information to traffic information centers through sensors and transmission equipment equipped on roads, in vehicles, at transfer stations, in parking lots, and at weather centers.[5]

When this feedback is received, the system is able to use big data technology for rapid processing, and the processed data plans are pushed out to traffic participants in real time. According to these plans, the traveler can clearly know the surrounding road traffic information, public transport information, transfer information, traffic weather information, and parking information, so that s/he can independently choose his or her own way of traveling, travel route, etc.

(2) **Traffic management information.** This system is mainly used by traffic managers to monitor the real-time traffic conditions, traffic accidents, weather conditions, and traffic environment, coupled with vehicle detection technology and big data processing technology and

[5]The content is quoted from Xia Huan, Editor in chief, *Collection of Data Centric Smart City Industry Solutions*. Wuhan: China University of Geosciences Press, 2016, p. 114.

it can have an accurate judgment of the current and future traffic conditions at a certain time. With these accurate judgments, it is much easier to control the traffic, such as the traffic signals, release of inducing information, road control, accident handling, and rescues.

(3) **Public transportation system.** Public transportation is the main artery of the city, and whether it is a bus or a metro, it must be safe, fast, punctual, and smooth. Through devices such as closed-circuit televisions installed on public transport and occasions, the traffic can be monitored in real time, and real-time information such as vehicle routes and stops can be pushed through multiple platforms. It is with this system that the urban public transport can operate more smoothly, otherwise urban traffic will become disorderly and chaotic, wasting material resources and time and polluting the environment (see Figure 2.3).

(4) **Vehicle control system.** The purpose of the vehicle control system is to help the driver control and drive the vehicle more efficiently and safely. The driver can be warned, helped, and commanded in case of an emergency. Many of the control systems installed on buses in our lives are part of this system.

(5) **Electronic tolling system.** A representative product of this system is ETC, the world's most advanced way of road and bridge tolling.

Figure 2.3 Mobile phone scanning code payment.

When the vehicle passes through the highway tollbooth, the equipment can quickly communicate and deduct money. ETC involves highway networking systems, bank-tolling charging systems and shortwave communications. The data generated by the interaction of these systems is an invaluable data resource. In addition to this, the current popular urban intelligent parking electronic tolling system also belongs to this. It is worth mentioning that Alipay's model of parking tolling without stopping is likely to completely overturn the parking tolling habits in the future.

(6) **Emergency rescue system.** Emergency rescue system is generally closely integrated with transportation information system and traffic management system. The other two systems provide timely feedback in the event of a traffic accident. After data analysis, the emergency rescue system can quickly provide emergency treatment, towing, on-site rescue, and troubleshooting of accident vehicles and other services for travelers.

Highway travel is the most frequent and most closely related mode of travel for travelers, so the areas and dimensions involved in the management process are also extremely extensive, and because of this, the data on highway travel is also the most complex and scattered.

2.4 Civil Aviation Travel of Big Data Transportation

In the "2018 Spring Festival Transportation Big Data Report" released by Tongcheng Travel and China Communications News, we can clearly see that civil aviation travel has gradually become the common choice of the people as they upgrade their consumption and change their travel methods.

For the 2018 Spring Festival transportation, railways were expected to transport 393 million passengers, up 10% from last year and well above highway transportation by 1.1%. As people's consumption perceptions change and civil aviation capacity increases, more and more consumers are choosing civil aviation during the Spring Festival transportation, which is expected to transport 67 million passengers in 2018, an increase of 14% over the same period last year, the largest increase among the three major transport sectors.[6]

[6]The content is quoted from the "2018 Spring Festival Transportation Big Data Report: More Rational Travel of People in Spring Festival Transportation, Huge Potential Demand for Intelligent Transportation Services", china.com.

For civil aviation travel, in addition to the fully online booking of tickets, seat selection for passenger, security, and baggage check have also been brought under comprehensive digitalization and intelligence. For civil aviation enterprises, the use of big data can not only further improve the level of service and reduce operating costs but also to ensure safety and good airport management.

2.4.1 *Big data of civil aviation is a treasure*

In civil aviation data, personal information holds an extremely large amount of data. This information includes booking information, payment history, flight process, high frequency routes, purchase preferences, transferee relationships, association effects, etc. Compared to the decentralized data of highway travel, civil aviation data are more centralized and is basically stored in the airline's ticket selling and check-in system, airmail seat booking and departure system, airport security check system, and other places. This information is integrated and optimized to accurately portray the customer's image. From the beginning of the booking process to the end of the journey, the customer's consumption habits and behavioral characteristics can be easily derived through this big data, and it is no longer difficult for airlines to personalize their services accordingly.

Moreover, civil aviation companies can expand their service offerings based on this information. For example, many airlines can cooperate with other companies, from exchanging miles for air tickets to exchanging commodities, from cooperating with ground and air intermodal transportation to sharing resources with airline partners, and to cross-border consumption points, all of which can help airlines explore the market and compete for differentiation, not only revitalizing their own assets but also enhancing the benefits (see Figure 2.4).

2.4.2 *Highly self-service of civil aviation*

Self-service travel has been developed in many areas of transportation, but relatively speaking, in civil aviation, in addition to security checks, other links also have high self-service, such as follows.

(1) **Self-service travel.** Starting from booking a ticket — passengers book their tickets in their smartphones and check road conditions,

Figure 2.4 Real-time traffic monitoring system.

flight status, and airport dynamics through their mobile APP on the way to the airport.

(2) **Self-check in.** Upon arrival at the airport, passengers check in first, and instead of waiting in long lines and they check in by themselves. For example, in Pudong Airport, the domestic airlines have made a breakthrough in the number of self-service check-in equipment, while in the past, domestic airlines uniformly provided less than 30 check-in equipment, but now there are 100. In addition, each airline has its own corporate culture and color characteristics, and Pudong Airport has also made some special personalized services in cooperation with the self-service check-in of major airlines — much more colors and much more efficient.[7]

(3) **Self-service luggage check-in.** In traditional luggage check-in, whether it is before take-off or after landing, passengers do not know the status of their luggage. With the popularity of big data, airlines can tell passengers about their luggage, such as whether it is already packed and where it is at the carousel when leaving the terminal. Passengers know the information well, so they do not have to bother the staff anymore, which greatly improves the efficiency of the airlines.

(4) **Self-service security check.** The security check traffic at the airport is not uniform. At peak times, when the number of people at security check is extremely more, it is especially important to divert the

[7]The content is quoted from "How does China's Second Busiest Airport Build a Full Process of Self-Service Travel", carnoc.com.

number of people at security check and get through the checkpoints quickly. Some airports have opened self-service ticket checking services, such as Pudong Airport, which has introduced a quick checkpoint service. After registering their ID, passengers will be able to take advantage of the favor that they will not have to wait in line. At peak times, there is a dedicated lane or a security check area for passengers, maximizing the time spent on passengers at security check.[8]

(5) **Self-service boarding.** In terms of self-service boarding, the self-service boarding system of Haikou Meilan International Airport was officially launched in 2016, thus achieving the first self-service boarding in China's airport area. Other airports in China are also making progress, such as Pudong Airport, which has opened self-service boarding at Terminal One. However, self-service boarding is still in the popularization phase, and some problems have arisen. Self-service boarding is not as efficient as it could be because passengers are still at the learning stage, and there are also issues of policy and security check regulations associated with self-service boarding, which needs to be promoted and adapted over time.

2.4.3 *Civil aviation big data can improve service quality*

In the past, the travel information, satisfaction, and demand of passengers were not truly integrated with civil aviation services, and could neither be docked at the same level nor could reasonable solutions be formed after data analysis. Now civil aviation big data can be completely up a notch in terms of service quality.

For example, TravelSky's Umetrip Software is the most authoritative and powerful civil aviation information service product in China, which can provide users with a full range of civil aviation travel information services and solve civil aviation travel problems through mobile phones. It has the characteristics of authoritative data, timely information, complete functions and comprehensive coverage, including real-time information of nearly 700,000 flights around the world, which can be automatically imported from the annual flight records according to user information, and at the same time, the software can clearly understand the

---

[8] *Ibid.*

reasons for flight delays and the implementation of the previous flight. Further, the software can also be used to implement the check-in through mobile phone, ticket search, ticket verification, ticket balance display, flight dynamics, airport information and boarding gate search, map navigation and find people in the same way and other functions.[9]

The continuous maturity of big data technology in the field of aviation will have great value in aviation information service and aviation operation efficiency.

2.4.4 *Civil aviation big data improves the aviation safety index*

Although the number of people choosing civil aviation travel has increased in recent years, many passengers have not developed the corresponding habits, exposing the uncivilized air travel behavior of Chinese passengers both at home and abroad. These uncivilized behaviors sometimes endanger the safety of other passengers. With big data control, these safety hazards can be eliminated.

For example, in 2015, the Bureau of Civil Aviation reached a consensus with the National Tourism Administration on working together to strengthen civilized tourism, curbing uncivilized travel practices and maintaining order in air transport and tourism. When uncivilized travel behavior occurs, the National Tourism Administration will share the relevant information with the Bureau of Civil Aviation, which will deal with it in accordance with the law and even add the related person to the "black list". If a passenger disrupts the aviation order or endangers the safety of others during the flight, the airline will handle the situation and share the relevant information with the National Tourism Administration to maintain a civilized travel environment.

In addition, for example, the flight big data of the aircraft itself will also be passed to the aircraft maintenance personnel of the airline as the relevant evaluation data to help them eliminate the potential safety hazards in time and ensure the flight safety.

[9]The content is quoted from "Big Data: The Gold Mine To Be Mined in Civil Aviation" by Ding Jian, Vice President of School of Computer Science and Technology, CAAC University. The content has been deleted and modified.

2.5 Freight Transportation of Big Data Transportation

Compared to passenger travel, freight logistics transportation and turnover are also the important parts of big data transportation and intelligent transportation. Moreover, with the improvement of vehicle networks and the Internet of Things, freight data are becoming more valuable.

From the point of view of the mode of freight transportation, the more important modes of transportation at present are rail transport, highway transport, waterway transport, and the civil aviation transport. The data generated by these freight transportation vary, but the key information is of common value.

According to the Statistical Bulletin on the Development of the Transport Industry in 2016 published by the Ministry of Transport in April 2017, the whole society had completed 43.134 billion tons of freight in 2016. Of these, the total amount of freight transported by railways nationwide was 3.332 billion tons, the amount of freight transported by highways nationwide was 33.413 billion tons, the amount of freight transported by waterways nationwide was 6.382 billion tons (excluding port data), and the amount of freight transported by civil aviation was 6.669 million tons (including international shipping data).

In a comprehensive comparison, the freight travel focuses on its flexibility, convenience, and densely populated outlets of highway transportation. In the process of construction and operation of intelligent transportation in the city, the highway transportation is more important. During the transportation process, the daily data volume generated by video surveillance, traffic police, road information, control information, operational information, GPS positioning information, and RFID identification information can reach the PB level and it is exponentially increasing. How these data are used wisely is at the heart of freight transportation of big data transportation.

2.5.1 *How to predict freight transportation in advance*

For freight drivers, it is the most cost-effective way to return with full load for every transportation. But in many cases, in order to transport goods, there is always a single transportation with no-load driving, which not only causes a waste of resources but also reduces the profit of freight drivers and increases the freight cost. If all the demand information is matched with the freight data, and the advance prediction is achieved, the freight drivers can reasonably arrange their own freight routes to ensure to transport the full load of goods.

In addition, there is also information asymmetry in the process of freight transportation. The shipper cannot find the appropriate vehicle to transport, and the freight vehicle cannot find the source of goods. Under the integration and scheduling of big data, such situations can be completely changed.

For example, the "Guan Che Bao" platform in the market is dedicated to solving such problems through big data. Through big data integration, it can dispatch millions of freight vehicle across the country to realize the aggregation and optimization of vehicle resources. In the process of transportation, the vehicle positioning can be inquired in the whole process, the driver's certificate can also be checked and verified, and the freight transportation is more guaranteed.

2.5.2 *Carrying out freight transportation plan and crossing peak transportation*

Freight transportation has an obvious characteristic of periodicity and concentration in time. According to some data institutes, the freight volume generally increases from Monday, peaks on Wednesday, and then declines until the end of the week. Time is extended to the whole month, which also has such characteristics. If there are major festivals or major events, the demand for freight transportation is very concentrated. According to the data precipitation in the process of freight transportation, big data technology can easily develop a freight transportation plan. After the information is provided to the owner and driver, it can guide the freight participants to plan ahead of time, cross peak transportation, and reduce the national traffic pressure.

2.5.3 *Freight data can better serve freight participants*

According to statistics, generally speaking, a large truck can drive up to 50 kilometers (including loading and unloading) in an hour, while most of the freight drivers start work at 7:00 a.m. every day and 11:00 a.m. is the time to deliver the goods, and 2:00 a.m. is the peak time because the large trucks are limited by the traffic management regulations of the city.[10] The research and integration of these data can clearly describe the various links and details of the freight participants. When the government management departments and relevant enterprises provide services, they will be more targeted, enhance the service quality, and improve the transportation efficiency.

[10]The content is quoted from Zhang Fan, Huang Jun, Su Zibo, Tian Chen, and Xu Chengzong's *Intelligent Transportation in the Era of Big Data (Long-Distance Freight Transportation)*. The content has been deleted and modified.

Chapter 3

Strategy Blueprint of Big Data Transportation

Looking around the world, many different countries have attached great importance to the development of big data transportation, especially Europe, the United States, Japan, and South Korea. Their big data transportation technologies are getting better by the day. In China's development process, only by continuously learning from advanced experience and introducing advanced technology can we develop our transportation network faster and better, improve the efficiency of people's travel, and reduce traffic pollution (see Figure 3.1).

Figure 3.1 Real-time traffic monitoring.

3.1 Big Data Transportation in the United States

Big data transportation in the United States developed earlier, and its technological sophistication and wealth of experience are far superior to other countries and regions. As the birthplace of global big data applications and the global center of big data development, coupled with the U.S. government's policy support for big data applications, as well as the comprehensively complete transportation network in the United States, the development of big data transportation in the United States leads the way for global big data transportation development.

From the logy of development, the United States adheres to the idea of government-guided norms and market-led development. The U.S. government has introduced a series of policies on big data research and applications, such as the release of the *Federal Big Data Research and Development Strategic Plan* in May 2016. These policies set the norms for enterprises to develop big data applications from a government framework. With the relevant norms in place, companies simply need to drive big data within the framework for industry convergence. As an important application area, big data transportation is bound to be vigorously developed by the government and the market.

It is with the support of policy and the associated technological base that the development of big data transportation in the United States has some distinct features.

3.1.1 *Years of technological sophistication and innovation*

Within the field of big data transportation, the years of technological sophistication in the United States have led to two developments. One is the accumulation and improvement of the transportation network. The U.S. has a dense transportation network and experienced infrastructure as well as governance in the transportation field, which is the foundation for big data transportation. The United States has also been implementing plans to establish an information superhighway since the 1990s. By now, the network infrastructure has been well established and protected as a strategic asset. Back in 2007, Walmart had built a mega data center with a storage capacity of over 4Pb.[1]

---

[1]The content is quoted from Liang Zhihao, Chinese Academy of Electronic Science, China's Enlightenment of Big Data Governance in the United States.

The second is the technical sophistication and accumulation of innovation in big data technology. As the world's big data center, the United States has a long history of convergence and application of big data technology in various fields. In addition, for many years, the U.S. government and enterprises have been working to gradually improve the relevant norms and regulations in the field of big data. More mature legal frameworks and ethical norms are already in place in areas such as data application, credit, and privacy. Also, in 2010, for example, the United States Congress passed an updated act to further increase the progress and frequency of data collection and reporting, making the data collection and aggregation system better at the national level.

3.1.2 *Extremely open data*

For the big data generated during the transportation process, not only does the government need the data for transportation management and transportation environment optimization but, more importantly, the companies also need the data to participate in every part of the transportation process so as to better formulate the corporate development strategies, optimize the customer service, develop the public service software, etc.

The United States has been implementing open government for a long time, including open data. The most important open data platform in the United States is Data.gov, which was launched in 2009 and is a key part of the United States' commitment to "open government". By November 2012, Data.gov had opened 3,88,529 raw and geographic pieces of data, covering about 50 categories such as agriculture, meteorology, finance, employment, demographics, education, health care, transportation, and energy, bringing together "everything from analysis of energy consumption trends in homes and enterprises to real-time global earthquake notifications, and even querying about the weather on Mars from data sent back by the Curiosity Mars Rover", according to the three categories of raw data, geographic data, and data tools. To facilitate public use and analysis, the Data.gov platform also incorporates new features such as hierarchical rating of data, advanced search, user communication, and interaction with social websites. The White House visitor search tool available on Data.gov, for example, can not only search for visitor information but also link White House visitors to Twitter and other social websites, further increasing the transparency of visitors.[2]

---

[2]The content is quoted from "Seven Strategies in Key Areas of Big Data R & D in the United States" on cnii.com. The content has been deleted and modified.

By opening up information in the field of transportation to the public, entrepreneurs and related enterprises can use the data to find business opportunities and provide more valuable public services to the public, thus making life easier and faster for the people.

3.1.3 *Numerous innovation elements and complete industry chain*

The transportation sector, which is closely related to people's lives, has a strong demand for its own industry. Moreover, the United States, as the world's largest country in terms of per capita vehicle ownership, has a greater demand for vehicles and transportation than other countries. It is because of this market that the United States has a cascade of startups in the field of big data transportation that are providing a continuous innovation impetus for the development of big data transportation in the United States. In particular, giant companies such as Google have a deep plough in the field of big data transportation, and the research and applications of new technologies such as driverless technology are built on this basis. These new technologies have further generated demand for public travel, driving enterprises, schools, and research institutions to input in it, and forming a complete industry chain.

There are many successful cases of the big data transportation in the United States. Let's look at a case of the U.S. Air Force using big data to dispatch vehicles.

The U.S. Air Force fleet supports the U.S. Air Force in flight and combat missions in the sky, space, and virtual space. In numerous missions around the world, it is required to equip a fleet of various vehicles, including buses, special, and military vehicles to complete complex missions.

As the fourth largest federal government fleet in the United States, with more than $7 billion in assets, the fleet is stationed at bases in the United States and around the world.

The U.S. Air Force bases at home and abroad are like "self-contained" cities, requiring a variety of vehicles, from cars to snowplows, fire trucks to fuel trucks, and more. Currently, more than 500 types of vehicles are authorized for use.

The U.S. Air Force fleet is governed by the U.S. Air Force Element Vehicle and Equipment Management Support Office, which sets vehicle regulations, subsidies, maintenance plans, prioritization plans, budgets, and more, and is based in Langley, Virginia.

The institute provides centralized vehicle management, secures 11 major combatant commands, 307 bases, employs 6,100 mechanics, and has an annual procurement program budget of $300 million. As the governing body, it proposes to safeguard actions in the most efficient way through continuous and rigorous process reengineering.

Its fleet management covers various aspects such as regulations, budgets, asset management, maintenance and repair, energy management, and regulation development.

Since 2001, the U.S. Air Force has been expanding its enterprise-level data warehouse, mainly using the Teradata Dynamic Enterprise Data Warehouse Platform 6690, which has now expanded to six nodes with 95.8 T of storage. It has also assisted in the deployment of the Informatics Power Center and related products from Business Objects.

From the data analysis evolution process of the U.S. Air Force, before 2007, it focused on the analyst view, and the original version only focused on one aircraft and its maintenance process. From 2007 to 2013, a rapid evolution from the enterprise view to enterprise interoperability was achieved. In 2009, the solution was extended to vehicles, equipment, and military supplies by supporting the addition of an asset query view, and in 2013, additional process managements such as inventory tracking, supply, and transport links were expanded.

Of these, the core components of the vehicle view application are divided into two parts. The first is the planning component, which includes authorization and validation, budget planning, and the priority procurement model. The second is the operational component, which includes the lead view, quick queries, cost of retention, high performance, and transaction request tools.

In authorization and validation, every mark on a U.S. Air Force base is verified, such as vehicle type, use agency, and use code, and each requirement will be repeatedly identified within 3 years. The workflow also makes it very cumbersome to create questions and regulations for the purpose of determining the need. Surveys are also generated for each base, and managers are expected to answer survey questions quickly. The results of this survey will serve as a rule-processing engine to provide insight into the calculation of vehicle demand.

In the face of the above, the U.S. Air Force has made management changes, for example, by calculating and comparing the vehicles that need to be authorized in the inventory at the moment. Actions are taken for each of the different users of the Air Force Element Vehicle and Equipment Management Support Office to conduct the approval or withdrawal of the application.

As can be seen in the changed summary analysis report, the statistics cover the increase and decrease in the number of authorizations, the different numbers of automatic and manual authorizations, etc. The final results show that the performance and approach have contributed to significant efficiency gains and cost savings. Statistically, the U.S. Air Force has saved a total of $309 million over the last 3 years through authorization and verification!

In the priority procurement model, it deploys the replacement strategy algorithm for the U.S. Air Force vehicles to efficiently plan vehicle procurement for the coming year, such as calculating the projected life cycle of each vehicle, giving priority to the replacement of each vehicle, allocating funds to different vehicle types to balance the fleet, simulating funding adjustments for hypothetical scenarios, analyzing fleet allocation metrics, and generating vehicle purchase lists for the upcoming fiscal year.

In the specific calculation, the life cycle of a vehicle means that when the cost of maintenance and operation is greater than the depreciated value, the vehicle is determined to have reached the end of its useful life. Each current asset of the same vehicle type is replaced in order of priority. Funding is determined by vehicle type, and each vehicle type receives an even distribution of funding within a category as a way to help with the balanced distribution of the fleet. The vehicle's replacement eligibility is that the user calculates the end of the vehicle's life cycle over a specified period of time (usually 3 years), and there is no vehicle allocation to replace it. This is followed by a calculation of the number of purchases, and the replacement of specific vehicles is based on the allocation of priorities. It also allows users to create hypothetical scenarios to see the impact of budget adjustments on model results, such as budget increases and budget decreases. An indicator of good vehicle allocation means not replacing a vehicle at the end of its life cycle, but using a structured, logical approach to measure the future impact.

The final result shows that in the first year of the model and the end of vehicle life structure, the procurement demand of air force vehicles is reduced by 1.5 million US dollars.

For example, the fleet's use of the Teradata geospatial feature is used to calculate the distance between bases.

Teradata's geospatial solutions provide fast and accurate analysis of geospatial data processing by integrating geospatial information into enterprise-grade data warehouses, and leveraging Teradata's powerful in-database analysis capabilities. At the same time, the integration of geospatial data with other business data in the data warehouse brings new data analysis functions to users.

For this reason, the U.S. Air Force fleet was named the "Top 100 Best Fleets" by Government Fleet Magazine, earning the honor of being a Green Fleet in 2014, as well as receiving 9 A4/7 annual and quarterly awards, 3 A4L awards, and the "Top 15 Federal Fleet Directors" award in the past 24 months.[3]

The development of big data transportation in the United States has been gradually perfected, involving not only government agencies but also market enterprises, in mutual integration and openness. In the future, it is bound to be a golden period of rapid development of big data transportation.

3.2 Big Data Transportation in Europe

In the development of big data transportation, in addition to the leading position of the United States, Europe, as a region with a highly developed transportation network, also has rich experience in big data transportation and its own unique development network. Furthermore, the European Union (EU) brings together European countries so that advanced development experiences of urban transportation can be shared and big data transportation technologies from each country can be disseminated.

The European city of Barcelona is the premier intelligent city in Spain. Its development in big data transportation and intelligent transportation is very representative of its status.

In terms of intelligent transportation, people in Barcelona can check the availability of rental bicycles at all times when they travel, just as people in China can check the distribution of common bicycles when they travel. In addition to promoting the popularity of rental bicycles, Barcelona has also promoted free charging stations, with more than 300 free charging stations for electric mobiles in the city.

[3]The content is quoted from Jiemian: "Decrypt How the U.S. Air Force Manages the Global Fleet with Big Data", which has been deleted and modified.

In order to make it easier for people to get around by public transport, Barcelona has upgraded the city's bus stops to smart bus shelters. The smart bus shelters are extremely intelligent, which is reflected in the following aspects:

(1) Smart touch system: Smart bus shelters are equipped with smart touch devices on which passengers can view relevant transportation information while waiting for a bus and use near-field communication technology and Quick Response "QR" codes to download multiple applications related to transportation, travel, and entertainment.
(2) Wi-Fi: There is plenty of Wi-Fi connectivity, as well as charging ports for smartphone charging.
(3) Smart advertising screen: The smart screen can push and display different advertisements based on the gender, age, and other characteristics of the waiting passengers.

In addition to the smart bus waiting hall, the Barcelona City Council has introduced a Smart Quick Pay system that allows people to pay for parking online when they travel. All people need to do is to register the relevant account and fill in the relevant information by self-service. When the parking is over, the system can automatically deduct the fee. On this basis, the City Council has invested heavily in the installation of a large number of sensors at parking spaces so that information about the parking spaces can be obtained in real time. Based on big data algorithms, the government's traffic management platform can push parking information to people in real time, helping them park quickly to reduce traffic pollution and wastage.

Moreover, there are smart traffic lights. The Barcelona City Council has provided small smart remote control devices that can be carried on the streets for blind people. When a blind person is crossing the street, the device will sound an alert when the traffic light turns green to inform him/her of the right time to safely cross the street. Besides, smart traffic lights can also be used with emergency access. For example, if there is a fire or traffic accident, when the emergency vehicle passes through the smart light, the smart light can turn the traffic signal green to let the emergency vehicle pass quickly. In this manner, the emergency vehicle can go all the way through to its destination.

This kind of intelligent transportation is very convenient for urban people to travel. Blind travel is reduced, which not only saves people's time but also reduces pollution.

Similarly, as a major EU member, Germany's actions in big data transportation are also very representative. Based on the perfect spatial geographic basic information framework of the EU and the advantages of the existing traditional information industry system in Germany, Germany's construction of digital cities such as big data transportation has been pushed to a very high level of development.

For example, Germany has a large number of sensors in the city, which can transmit the data of traffic, weather, and pedestrian movement to the data center. The data center will inform the City Council and the road administration bureau of relevant information, and the relevant government agencies will set the traffic equipment such as streetlights such that the brightness of the streetlights can be automatically adjusted according to the actual needs.

Further, in the early stages of building big data transportation and intelligent cities in Germany, the construction of information infrastructure was the main focus. The most important of these is the construction of basic geospatial information, which is the carrier of natural, social, economic, humanistic, and environmental information, and is the basis of big data transportation and intelligent cities. According to relevant information, German Länder law requires all information systems to use basic topographic maps (official) provided by the Länder surveying and mapping departments. The coordinate system of the topographic map is unified throughout Germany, as are the standards for the collection and storage of topographic geographic information and the collection and storage of information on the transportation network.

In order to facilitate the management and sharing of data and their use, Germany has adopted a uniform data exchange format through consultations of the Coordinating Working Committee. Data collected from all sources can be exchanged through standard data exchange software to become a data format that can be freely used by all groups and industries. In this way, the sharing and use of big data become very convenient, avoiding duplication of construction and waste in the collection, storage, and use of data resources.

We can see that Europe has its own unique advantages in the construction of big data, which can be summarized as follows.

3.2.1 *The development of big data transportation is more humane and more beneficial to the people*

No matter whether the UK, France, Germany, or Spain, European countries have always taken people's travel needs as their main objective in the construction of big data transportation. Moreover, the government has invested heavily in the construction of the relevant infrastructure, which is convenient not only for the population but also for European enterprises. The government and enterprises have cooperated to build a large amount of big data transportation infrastructure, which provides good conditions for the development of big data transportation.

Many countries and regions, including China, have not done enough in standardization during construction of big data transportation. Regions and administrative divisions always set their own standards, making the collection, storage, and use of big data on transportation fragmented and unable to form a unified call. Even when countries unify to call through administrative means, there is often a significant waste of resources due to data format problems. Some emerging countries, in particular, are starting out in the construction of big data transportation, and if they do not do a good job in standardization, it may lead to a lot of waste of resources and duplication of construction.

3.2.2 *Financial investment is increased*

The construction of big data transportation often requires the government to coordinate and guide the unified formulation of relevant policies and norms. If local administrative units or enterprises are allowed to develop, it often results in information silos where governments or enterprises, even if they have huge data resources, cannot integrate with other data resources, and finally waste them. This would have been effectively avoided if the Government had invested heavily and uniformly in the early stages of construction.

3.2.3 *The construction of big data transportation is focused on public transportation*

No matter how diverse people's travel needs may be, as a whole, it is public transportation that is needed by the majority of the population. Especially in the construction of intelligent cities, public transportation is the most important. The construction of big data in public transportation requires not only strong government investment but also that the government and enterprises educate the public, so that the public can also take the initiative to participate in the construction of intelligent cities.

As the leader of big data transportation and intelligent city, Europe's construction experience and achievements are worthy of our careful study. In this way, China's construction of big data transportation can be more characteristic and intelligent.

3.3 Big Data Transportation in Japan

We have already mentioned above that as early as 1990, the Japanese Masaichi Iguchi first proposed the concept of Intelligent Transportation System (ITS for short). Although such a concept was later promoted by the Americans, Japan has never lagged behind in the construction of intelligent transportation and big data transportation. Among the Asian countries, Japan is ahead of other countries in the construction of big data transportation, and it is worth learning from.

In 2013, the Government of Japan issued a document entitled "Declaration on the Creation of a State-of-the-Art IT Nation". In this document, the Government of Japan comprehensively elaborated its national strategy for the period 2013–2020, which focused on the development of open public data and big data, and emphasized that "the application of big data is indispensable for enhancing Japan's competitiveness". Such a government document is a reflection of the Government of Japan's investment and determination in big data construction and big data transportation applications. With government support and regulation, the development of big data transportation will naturally be more vigorous.

Japan's construction of big data transportation has performed extremely well in preventing traffic accidents and ensuring traffic safety.

In the 1990s, Japan began to promote the construction of data transportation and Internet of Vehicles. Since 1995, the number of traffic deaths in Japan has been decreasing because of the construction of data transportation by the government. By 2013, the number of traffic death in Japan had fallen to 4,373, which is already an extremely low death rate for accidents.

In Japan, there are two general categories of safety services for the Internet of Vehicles, namely, hazard information provision and accident emergency assistance. Hazard information provision refers to the sending of traffic information, such as accident-prone hazard information and weather conditions, to vehicle users through vehicle-to-vehicle information exchange (referred to as V2X) and mobile Internet. These data are produced through the mining and analysis of big data, which can accurately reflect the traffic hazards and warn vehicle users to pay attention. The main purpose of hazard information provision is, of course, prevention, so that traffic participants are aware of prevention and can take precautions.

Emergency accident rescue, on the contrary, focuses on the guarantee of time and efficiency, so that after a traffic accident, rescue can be carried out with the greatest efficiency thereby reducing the death rate. For example, in Toyota T-Connect's emergency rescue, when an accident occurs, vehicle users can contact the backstage staff through HelpNet with one click, and with the GPS information, rescuers and police will quickly arrive at the site of the accident. When the accident is serious, the Internet of Vehicles can directly contact the backstage without requiring anyone to press the button. Such a system would then be able to protect the traffic participants to the maximum extent possible and reduce losses.

As we all know, the industry chain of the Internet of Vehicles can be divided into car manufacturers, TSP, onboard terminal suppliers, network operators, and other links, and its function is nothing but safety, driving, entertainment, and services. At any link, there is a huge amount of data generated, which may be needed by government administration or by the enterprises involved. Honda Motor Company's Internavi can calculate the blocked roads based on VICS (Vehicle Information and Communication System) information from Japan, FCD (Floating Car Data) stored inside Honda Motor, and traffic congestion statistics. This road information is what the government wants because, combined with its own transportation system, it can analyze the causes of traffic congestion and improve traffic conditions. By sharing the information from Honda's Internet of Vehicles with the government, the government can reduce some of the repetitive construction and spending. For Honda, with these data, when providing information services to users, it will be possible to guide them precisely to avoid blocked roads and reach their destination more quickly. Many map navigation software and car companies are now working together to share information and improve service efficiency, which can also reduce fuel consumption and environmental pollution.

According to research, Honda's Internavi environmental route service is based on calculating the fuel consumption of each route through

Honda's internal stored FCD, and then accumulating it at a set destination, so as to recommend an environmentally friendly route with the lowest consumption. For example, from Yokohama City to Fuchu City, Tokyo, the eco-friendly route takes 5 minutes longer but reduces fuel consumption by 0.8 liters and CO_2 emissions by 17.8%. In addition to this, B2C applications for the Internet of Vehicles also include onboard remote control, maintenance reminders, and smart parking.[4]

Furthermore, Japan's Internet of Vehicles is used in conjunction with external data. For example, VICS real-time information in Japan is mainly pushed by roadside devices on high-speed and arterial roads, but there are sections of road without roadside devices, so the system can only rely on static information on the map to determine the road information. When plugged into Honda's FCD, road information can be captured in real time. The combination of Internavi and Honda's FCD enables the vehicles to receive traffic information pushed from the Internavi information center even when the vehicle is traveling on a road without roadside devices. The same effect can be achieved with high-speed and arterial roads, greatly expanding the road range of intelligent transportation.

Japan's intelligent transportation and construction of big data transportation combine the strengths of the United States and Europe to form big data transportation with its own characteristics as shown below.

3.3.1 *Government-led and unified norms*

The Government of Japan has been playing an active role as a facilitator and financial contributor in the construction of big data transportation. It is because of the investment in infrastructure and the unified norms and framework that the development of big data transportation and the Internet of Vehicles in Japan has been very strong, especially the development of the Internet of Vehicles, which has its own characteristics.

3.3.2 *The car companies are highly electronic*

As an important player in big data transportation, car companies compete with the U.S. and Europe in the market, in order to further strengthen their

[4]The content is quoted from "New Opportunities Brought by the Internet of Vehicles, Starting From the Experience of Japan", which has been deleted and modified.

competitive advantage and constantly improve their own electronic level. Automotive giants such as Toyota, Honda, and Nissan have been actively cooperating with electronics manufacturers since the 1980s and have jointly developed devices such as vehicle body electronic control and onboard electronic information to improve the electronic level of vehicles in Japan. Such a means invisibly lays the foundation for the development of big data transportation. With electronic devices, the collection and storage of data information becomes easier, and in the construction of big data transportation, the investment can be reduced and the time can be shortened.

3.3.3 *High degree of marketization*

In the development of big data transportation, governments in some countries and regions have played greater roles, while enterprises in some countries and regions have a higher level of participation. In Japan's development of big data transportation, automotive industry giants have partnered with electronics giants to facilitate the immediate transfer of data and the real-time recall and push of data resources.

It should be noted that since 2007, the OBD2 system has been compulsorily introduced in all cars manufactured in Japan. The system can automatically detect problematic faults in equipment such as lights, engine, and heat exchangers. It can also monitor exhaust sensors, tire pressure, battery status of electric vehicles, and read out vehicle data such as speed, fuel consumption, and mileage. By downloading the corresponding APP on the user's mobile phone, the user can get the vehicle information and other information in real time. Governments and enterprises can provide a better service if they integrate this information. For example, newly opened roads are automatically drawn on the map based on the location information of the FCD, and accident-prone areas are identified based on the statistics of emergency braking and the user is alerted by voice. By linking Honda's Internavi information center with the Japan Meteorological Society's database and by linking the FCD with real-time weather information, Honda's services to users will be more accurate and personalized.

Car enterprises play an important role in the construction process of the Internet of Vehicles and big data transportation. There are many things that the government does not do, and car enterprises inadvertently do things that the government does not do in order to cope with competition.

3.3.4 *Information sharing is adequate*

Japanese companies and the government are doing a better job of sharing and using transportation data. The enterprises can share various data resources with the government, and the government can also provide better services for the enterprises based on the data resources collected and stored by it, so that the enterprises can reduce the cost of data use and will be more competitive in the face of international competition.

In conclusion, Japan's big data transportation development started very early and is well developed. With the development of technologies such as cloud services and artificial intelligence, the data resources available to Japanese car companies in these areas will be even richer. It would be a win–win situation for governments with huge transportation resources to share their resources more rationally and fully with companies.

3.4 Big Data Transportation in China

The development of big data transportation in China is in a period of rapid growth or at its beginning. Although China's transportation network has seen great development in recent years, platforms for access to and storage and use of transportation information have emerged, and the government is investing heavily, there are still many shortcomings.

The advantages of China's big data transportation development are extremely obvious, which can be found from the following aspects.

3.4.1 *The development has been rapid*

The speed of China's economic development is evident to the world. Since the beginning of the 20th century, with the popularization of the network, big data transportation in China has begun to expand rapidly. The growth of infrastructures such as transportation facilities and routes is the foundation, while the highly centralized generation of data resources is the top priority, and the amount of data resources is also rapidly expanding. According to relevant statistics, the data volume of big data transportation in China will double every 2 years, which is a very daunting data increment. In the process of transportation, with the increase of the number of bayonets, electronic police, cameras, sensors, and other equipment, as well as the growth of Internet data transmission speed, there will be blowout development in the field of big data transportation in China in the future.

For example, Guiyang, China's Big Data Center, launched the Transportation Big Data Center in 2016.

The first phase of the traffic big data center will integrate the internal data of the Municipal Transportation Commission, including the integration of six types of data low-carbon data center, data cage, bus data, methanol car data, and water search and rescue. The second phase will integrate horizontally coordinated data, including data from the provincial transportation department, housing construction data, meteorological data, traffic control data, city control data, parking data, railway data, and civil aviation data. The third phase will report to the municipal government to coordinate the data of Sinopec and PetroChina, such as the signaling data of intelligent Internet-connected vehicles and mobile operators' mobile phones, and the data of new energy vehicles, so as to realize the "all in one network" of the traffic elements.

The completion of the traffic big data center will realize the visualization and comprehensive analysis of public transportation, comprehensive scheduling, congestion index, vehicle demand, and construction planning. Big data technology is used to carry out quantitative analysis of the characteristics of urban passenger transportation, to explore the city's transportation demand hotspots, distribution laws, changes in characteristics, etc. The self-processed early warning, precognition, and forecasting in the areas of traffic congestion, resource allocation, and weak law enforcement are formed, and a new management model of intelligent transportation to guide scientific decision-making and intelligent scheduling in the industry is built.

In the future, Guiyang will, based on the mechanism of overall command, coordination, and coordinated dispatching of the Municipal Transportation Commission, connect the provincial, municipal, and county traffic data service resources vertically, integrate the public service information of water, land, air, and port horizontally, and form "one command" for dispatching, complaint, and supervision.

Guiyang will integrate the national traffic complaint service hotline 12328 and mayor's hotline 12345 to form an external public service function integrating passenger transport hub, public transport service, taxi complaint, and law enforcement management, as well as the early warning and forecast of urban public traffic congestion, traffic service event supervision, etc. It will complete comprehensive coordination with the public security and traffic control law enforcement force and the joint law enforcement of fighting against non-compliance. It will also coordinate with aviation and railways to form data fusion in connection transportation, transfer service, and other aspects, and promote the "zero transfer" of a comprehensive transport hub in Guiyang.[5]

3.4.2 *Data types are diverse and data resources are abundant*

China's traffic situation is relatively responsible, including highway, railway, shipping, and other modes of transportation, which will produce a large number of data resources. As far as highways are concerned, urban roads, expressways, township roads, and other road sections generate a large number of data resources. The main types of these resources include sensor data (location, temperature, pressure, image, speed, RFID, and other information), system data (log, equipment record, MIBs, and so on), service data (charging information, Internet service, and other information), and application data (manufacturer, energy, transportation, performance, compatibility, and other information). Therefore, there are many types of transportation data in China, and the volume is huge. In the future, as the population moves from the countryside to the city, more and more data such as video monitoring, bayonet alarm, road condition information, management and control information, operation information, GPS positioning information, and the RFID identification information will appear in the traffic section. These data are vast and precious.

Diversified data can provide corresponding support for government departments and enterprises. For example, we can deeply mine big data transportation, and the public security department can launch big data models such as vehicle trajectory, road traffic, and case clustering. Based on the big data model, common functions can be launched, such as recognition for license plate, vehicle body color, vehicle model, vehicle logo, model year, and other features, and analysis for sunshade detection, seatbelt detection, call reception detection, and driver face recognition, which can be used for intelligent car following analysis, track collision, face comparison, public opinion analysis, etc. Such data resource exploration can help government departments gradually solve the deep-rooted problems of the industry.

---

[5]The content is quoted from the Transportation Big Data Center Launched in Guiyang on jt-w12345, which has been deleted and modified.

In addition, the government departments can use such big data transportation management systems to obtain road weather, construction situation, and accident situation data, and combined with big data analysis, it can also provide information such as weather, road condition, accident-prone location, and parking lot for drivers and traffic management departments. Relevant service enterprises can recommend driving routes according to vehicle destination, driving habits, and road conditions.

3.4.3 *The advantage of late development is obvious*

The big data transportation technology and experience of the United States, Europe, and Japan have been accumulating for many years, and China has more backward advantage to develop on the advanced foundation of these countries and regions. On the one hand, China's government and enterprises can learn from the existing construction experience and technology, rapid transformation, and use of big data transportation. On the other hand, the government and enterprises can innovate and localize on this basis. For example, China's high-speed railway, one of the "four new inventions". is a high-speed railway product with independent intellectual property rights based on absorption and introduction.

However, compared with the development of big data transportation in other countries and regions, there are still many problems in the development of big data transportation in China, which need to be faced and solved jointly by the government and enterprises.

3.4.4 *Inadequate integrated planning*

China has special national conditions, and the development of each province, city, and autonomous region has different conditions; so, in the construction process of big data transportation, there is too much local color at present. Each administrative region develops independently and does not form a situation of national unity. Therefore, there is a lack of national unified planning in the collection standard, storage standard, use specification, and sharing principle of transportation data resources. At the early stage of the development of the big data transportation, the situation may be vigorous and rapid. However, with the formation of the transportation network connecting the whole country, the transportation connections across the country are closer and closer, which leads to a bottleneck in the development of big data transportation without unified planning before, and it is easy to form an information island.

3.4.5 *The division of the functions of the government and administrative departments has led to the failure of closed loop management*

When it comes to transportation, there are many administrative units in China, such as the Ministry of Public Security, the Ministry of Transportation, and the Ministry of Construction. These departments, in the process of transportation management, will miscommunicate, and with each doing its own thing, it creates a kind of managerial confusion with each department having its own requirements and standards.

If we can learn from the big data transportation development in Europe and Japan, unify and coordinate government departments, scientific research institutions, enterprises, and other relevant subjects, divide the scope of their respective responsibilities, and carry out unified planning, we can each give full play to its advantages and focus on doing what we are best at. For example, government departments focus on planning, departmental coordination, policy research, technology development, standardization, market order maintenance, quality supervision, information services, etc. Enterprises can then concentrate on developing a variety of public service products that are as convenient as possible for transport participants.

3.4.6 *The financial investment should be balanced*

The government will continue to invest in the process of building big data transportation on an ongoing basis. However, the development of each region is different, which leads to the government's unconscious focus on some regions or aspects of capital investment. Although such financial investment is beneficial to some regions, it will lead to the failure to form a unified and valuable big data transportation network in China. In addition, the imbalance of financial investment may cause some technologies in the field of big data transportation to lag far behind others and finally be controlled by others.

3.4.7 *Insufficient sharing of data resources*

With the change in policy, the data resources of the Ministry of Public Security, the Ministry of Transportation, the Ministry of Construction, and other relevant administrations are in the process of gradually opening up. However, it is still difficult for enterprises to access these resources. China has not formed a unified data sharing platform and also lacks relevant data resource security protection, resulting in a lack of data security for both government departments and enterprises. Still, government departments and companies are scrupulous about sharing data resources. However, if information sharing is not possible, information islands will be formed, and even if data resources are abundant, their value will not be reflected. In this regard, we should learn more from Japan's experience, and the government and enterprises should work closely together.

3.4.8 *Lack of industry talent*

The development of big data transportation is rapid, but the related talent training is lagging behind. There is a shortage of talent in big data analysis and key technologies in transportation, in particular, key technical talents in the field of transportation; if they cannot cultivate themselves, they will be constrained by others for a long time, limiting the development of big data transportation. These dimensions require integrated government planning and coordination.

Chapter 4

Big Data Transportation: Eight Innovative Modes

Big data transportation has been gradually popularized all over the world, and many representative cases have emerged in this process. These innovative cases, whether successful or unsuccessful, are worth learning from by government decision makers and entrepreneurs.

4.1 Feature 1 of Big Data Transportation: Shared Economy

Although the shared economy model has developed relatively early abroad, its development speed and innovation models enhanced only after entering China. Through the mobile Internet and big data, the shared economy can reasonably reallocate resources, improve the utilization rate of resources, and reduce the cost of ownership. It also reduces waste and has obvious effects on the improvement of economic benefits to the community as a whole.

The shared economy has several characteristics:

(a) The allocation of resource elements is rapid and efficient.
(b) Ownership changes and proprietorship and usage rights are separated.
(c) All resource allocations are based on data mining from shared platforms.
(d) Everyone can be a participant in the shared economy.

(e) User experience is critical and participants' behavior data are presented on the platform, and participants' behavior is governed by evaluation mechanisms.
(f) Cultural attribute changes, individual ownership does not mean everything, and only when individuals use it will they reflect the true value.

From these characteristics, it is easy to see that the core value of the shared economy, regardless of the field, remains resource allocation and user experience, which requires big data to do so. Within the transportation sector, there are many examples of entrepreneurship under the sharing economy model, the core elements of which are still resource allocation and user experience.

According to the *European Times* of January 18, 2017, BOLLORE that designs and operates the Paris self-service electric car rental system (AUTOLIB'), has filed a financial paper stating that the AUTOLIB' posted a loss of €179 million.

Paris electric car rental system was launched in October 2011, using BLUECARS electric cars designed and operated by the BOLLORE Group. Today, there are electric car rental stations in the Paris metropolitan area and in the suburbs of Paris. The service is very popular, as it currently has 130,000 subscribers. When the system was launched, BOLLORE Group said that it would become profitable when it reached 100,000 subscribers.

However, the BOLLORE Group claimed that AUTOLIB' was spending much more than expected, which was regularly losing money and was a bottomless pit. According to figures published by *Le Canard enchaîné*, financial documents handed over by the BOLLORE Group to the City of Paris stated that AUTOLIB' posted a loss of €179 million.[1]

For Autolib', it has the strong support from the city of Paris. A self-service rental system has been adopted, the core element of which is the analysis and mining of big data to make it as easy as possible for citizens to travel, and improve the efficiency of resource allocation. However, because of the problems of the operating company, it ended up being a failure.

In comparison, China's shared bicycle is much more successful. The domestic shared bicycles represented by Mobike and OFO Little Yellow Car used to be a dockless shared bicycle travel platform in China. It created

[1]The content is quoted from Zhou Wenyi, "AUTOLIB's Loss of Nearly 200 Million and the Paris Government Refused To Pay For It", *European Times* (Chinese Version). The content has been deleted and modified.

a "doseless bicycle sharing" model and is committed to solving urban travel problems. Users only need to scan the QR code on the car on the WeChat official account or App or directly enter the corresponding license plate number to get the unlock password, unlock the ride, take it and use it anytime, anywhere, or share their bike to OFO sharing platform, get the lifetime free use right of all OFO small yellow cars, exchange 1 for N. Since its launch in June 2015, OFO's small yellow car has connected 10 million shared bicycles, and has provided more than 4 billion trips to more than 200 million users in more than 250 cities in 20 countries around the world. From October to November 2018, OFO was included in a number of cases in Beijing No. 1 Intermediate People's Court, Beijing Haidian District People's Court and many other courts, involving 53.60 million yuan of execution exceeding the limit. At present, OFO shared bicycle brand has withdrawn from the Chinese market. OFO are developing rapidly. Since the birth of the shared bicycles, in more than a year, the scale of active users has reached as many as 70 million, and the loyalty of users is also relatively high. Especially in the demand of short-distance travel, the frequency of using shared bicycles is very high.

Through the combination of user location information and bicycle location information, the shared bicycle can reasonably allocate the distribution location of bicycles, allowing users to quickly find the bicycles. When a user uses smartphone to scan to unlock a bicycle, the smart lock will authenticate the user's information and then open through a signal connection between the smart lock and the bicycle's operational backstage. The entire cycling process, the user's cycling track, journey, time, and other data can be seen in the background of the shared bicycle. These backstage big data are not only used by bicycle operating companies to dispatch bicycles to meet user demand, but also for precision marketing.

According to the *2017 White Paper on Shared bicycle and Urban Development* released by Mobike, the total distance ridden by shared bicycles in China has been more than 2.5 billion kilometers, which is about 3,300 round trips from Earth to the moon, and cycling is a relative substitute for some private car travel, which brings considerable "green effect". The birth of the shared bicycle has doubled the use of bicycles for travel. In cities, the shared bicycle has become a good alternative to public transportation (bus, metro) and private cars. To a certain extent, the use of the shared bicycles has reduced the dependence on private cars, and the proportion of private car travel has decreased by about 3.2% after the emergence of shared bicycles.

The shared bicycle is just a microcosm of the shared economy in transportation. With the shared car, shared motorcycle, and shared bus popping up all over the place, these sharing models have an impact on big data transportation in the following ways.

(a) It revitalizes the existing transportation resources and increases the utilization of resources. The shared car, such as Bao Jia Car Rental, has revitalized private cars and this has increased the use of resources for the whole society (see Figure 4.1).
(b) It enhances the efficiency of travel. For example, the shared bicycle solves the problem of "the last kilometer of travel", and in short-distance travel, it changes the past situation of illegal motorcycle-taxi, making people travel freely and thereby reducing traffic congestion.
(c) Data analytics makes transportation smarter and more purposeful. Under the shared economy, the users have a strong purpose when using transportation, together with the precise solution under big data mining, so that the people do not have the blindness of previous travel, and the traffic congestion can be alleviated.
(d) The shared economy model makes urban transportation smarter and more environment friendly.

Figure 4.1 Electric cars charging at the roadside.

Source: www.DIEV.com.

4.2 Feature 2 of Big Data Transportation: Full Utilization of Resources

The high vacancy rate of cars in traffic has always been a headache for traffic managers. A vehicle with five seat, when fully seated, is fine for five people to share one vehicle. If five people each drive one car, there will be four more cars on the road, and these cars will naturally increase traffic congestion and the exhausted fumes from the cars will also increase pollution.

The mode that allows five people to share a car is the hitchhiking mode or carpooling mode. Didi Taxi has long offered hitchhiking and carpooling services. When a passenger chooses to carpool, a Didi Taxi driver can pick up and drop off several passengers to different destinations at the same time, thus reducing the number of transport used to travel and also ensuring that passengers are able to travel normally with traffic congestion eased considerably.

According to the statistics, the Didi hitchhiking service transported 4.2 million passengers during the 2017 Spring Festival, which is equivalent to opening 1909 green trains or 5874 8-car EMUs on the basis of the established railway transportation capacity. This is four times the number of trains transported in the same period last year. A total of 2.8 million car owners participated in empty seat sharing. Even on New Year's Eve (from 22:00 on January 27 to 6:00 on the next day), 70,000 people left for the reunion on their way home. During the Spring Festival (January 28–February 1), cross-city hitchhiking is mostly short-distance trips, family visits, and short trips in surrounding cities.

In terms of the number of passengers, Wenzhou people like to travel alone the most, and 66.3% of Wenzhou people hitch a ride alone on their way back. Guangzhou people like to travel with friends the most, and 52.11% of Guangzhou people hitch a ride together when they return.[2]

In the modes of carpooling and hitchhiking, idle resources are used to the greatest extent, which makes these modes an effective way to relieve traffic pressure. The most important thing is that the government and enterprises get the real first-hand data from people's travel through the hitchhiking and carpooling. From these data, the government will formulate its traffic relief plan and decision-making, and the enterprise's service will be more accurate.

---

[2]The content is quoted from Didi Hitchhiking Big Data, "People Are Shocked To See This Number on Bohai Morning Post". The content has been deleted and modified.

In addition to carpooling in small cars, other modes of transportation have also experimented with carpooling. For example, in October 2017, the Xi'an Railway Bureau conducted train carpooling through crowdfunding, and passengers who participated in the crowdfunding took the train on time at a predetermined time. The train changed its original departure time and replaced it with a time that was convenient for passengers. In this way, the passengers will not encounter a situation where they cannot take a bus or taxi late at night once they reach their destination. This attempt is another form of carpooling, but it is expressed in the form of crowdsourcing. If developed to maturity, many special trains could be customized by carpooling to avoid empty trains (see Figure 4.2).

Of course, buses such as the Dada Bus also have carpooling and hitchhiking modes. Regardless of the mode, the data left on the platform are the most real and valuable. It is with the support of these data that big data transportation is more efficient and intelligent.

4.3 Feature 3 of Big Data Transportation: Precise Demand

Precisely matching demands and providing services with precision are vital characteristics of big data. In the transportation sector, in order to be precise, both the service provider and the people served must have precise data, and demand and service can be perfectly matched. When the

Figure 4.2 Highway at night.

data is large and rich enough, quantified demands will come to the fore. At this point, it is not difficult to provide precise demand services.

As a big data center in China, Guizhou is at the forefront of precise big data services within the transportation sector.

Since the launch of the big data fusion platform with "data cage" type, Guiyang Transport Administration Bureau has gradually realized the real-time monitoring and full supervision of administrative power operation platforms such as administrative law enforcement, administrative approval, and public service.

The blind spots of public transport services such as residential areas, industrial parks and schools with more than 30,000 people are the key service areas of "customized buses". Customized public transport is aimed at the relatively concentrated area to recruit and reserve passengers, and the customized service of point-to-point and fixed seat for fixed person is implemented.

Guizhou Qianxi Green Taxi Service Co., Ltd. was the first company in China to obtain the operating license for online reservation of taxis, which marks the first brand of tailored taxi service in China to operate legally and opens a new chapter in Guiyang city's transportation data.

Qianxi Company adopts the B2C operation model of "professional vehicle and professional driver" to provide every passenger with a safe, comfortable, and convenient travel experience.

At 5 p.m., Chen Daiyu, who works near Lewan International, gets on the bus to go home. "Everyone has a seat, so close your eyes and relax yourself, and you'll be home soon", Chen Deyu said with a smile. But before July last year, when Chen Daiyu was ready to take a bus home from work, he was not in such a relaxed mood. "At that time, it was very troublesome to wait for the bus, because the time was uncertain. It took a lot of time if missed a bus. The taxi in this place was not easy to take, and the price was expensive, so at that time, as long as I took a bus from work, I was anxious".

In July last year (2015), the first "customized bus" from the Provincial Museum on Beijing Road to Wudang Lewan International opened in Guiyang. Many office workers like Chen Daiyu who need to travel between the two locations felt greatly relieved. "After the opening of the customized bus, it was on time at a fixed time, and is in one-stop direct from home to work, saving time and improving the quality of work and life". Subsequently, Guiyang Public Transportation (Group) Co., Ltd. launched four "customized buses" from the provincial government to the Science City, the city's administrative center to the Memorial Tower, etc.

In addition, the general public too began to enjoy the convenience of "customized buses", like students in Guizhou University City this 2015 year also began to take the "customized bus" on weekends and holidays.

In April 2016, the first class of the "customized bus" opened in the University City, and hundreds of students arrived the Jinyang Bus Station and Guiyang Railway Station safely and quickly by "customized buses". "Taking the 'customized bus' means we will not have to worry about not being able to squeeze into the bus, which will eventually lead to not being able to get on the train back home", University student Xiaoli said.

"'Customized public transportation' is one of our 'customized buses', which is also the earliest trial project. At present, from the perspective of the response of the public, it is quite acceptable and has accumulated experience for us to implement 'customized bus' in the whole city as the next step", Liu Xianglong, deputy head of Passenger Transport Department of Guiyang Road Transportation Administration, said.

Liu Xianglong told reporters that the customized bus is an important part of public transportation. Based on the civil use of big data transportation, the municipal transportation administration has built a customized bus network service platform. The operation supervision and application services of the platform are integrated into the overall management of Guiyang Low Carbon Transportation Information Command Center. The data center is built by relying on Guiyang data resource center, and the relevant data are connected to the data exchange and sharing platform of Guiyang municipal government.

Through the Internet and back-end data processing and analysis, the platform provides basic functions such as market survey, demand analysis, route pushing, price calculation, frequency booking, network payment, publicity and announcement, passenger interaction, vehicle management, group booking, and campus service of customized bus.

Guiyang Transportation Bureau's customized bus network service platform has been built since last year, and it has been basically completed and will be put into use soon. "The blind spots of public transport services such as residential areas, industrial parks and schools with more than 30,000 people are the key service areas of 'customized buses'. In addition, 'customized bus' can also make up for the lack of public transport capacity and meet the personalized travel needs of the masses".

The platform calculates the optimal route based on big data and users' individual needs, and the travel time can be shortened on average compared to traditional public transportation, solving the three pain points by providing "uncrowded, inexpensive and fast", service and making it an effective means to ease traffic congestion and improve the travel experience during peak hours.

"Self-processing" data push enhances the precision of law enforcement.

On May 18, Xia Xungui, the driving training industry special administrator of Auditing No. 2 Brigade in Guiyang Transportation Bureau, constantly kept receiving warning signs of "data cage" in his mobile APP. "The reason is that in a driving training school in Yunyan District a coach had left a car beyond the 'electronic fence' range for three days". Xia Xungui told reporters, at the end of 2015, that Guiyang Transportation Bureau has launched a new timing system for the training of the drivers, which became one of the six sub-modules of the performance efficiency of the big data fusion platform of the Bureau. One of the important functions of the system is to monitor the training sites of the instructional cars, forming an "electronic fence" that can track the driving status of the instructional cars through positioning systems.

"In the past, it was completely impossible for us to supervise thousands of instructional cars in the city. Now with the big data fusion platform, real-time management of instructional cars through big data is very convenient and effective. And the platform found anomalies will also start self-processing and active warning", Xia Xungui said.

On May 19, 2016, Xia Xungui came to this driving school to verify the operation of the training car for platform warning. "It makes our enforcement more accurate and effective", Xia Xungui said.

In addition to the instructional cars sending of an alarm beyond the electronic fence, the big data fusion platform of Guiyang Transportation Bureau can also initiate self-processing alerts for all types of abnormal data, such as vehicle operating right expiration and rating expiration. "Each type of abnormal data is judged by different criteria, and abnormal data must be processed in a timely manner if found. If it is not dealt with within the specified time, it will be reported to the supervisors".

Big data fusion platform will also integrate relevant data, and comprehensively monitor the workflow of various departments. Once there is abnormal data and irregular process operation, the platform will be alarmed. Staff should conduct self-examination within three days. If there is no reasonable explanation of the situation, the platform will automatically push the abnormal situation to the leadership in charge after three days. If similar anomalies occur frequently within a certain period of time, the Bureau's discipline inspection and supervision department will be involved in the investigation.

Rui Jin, Director of Guiyang Road Transportation Bureau, said that the Bureau will continue to promote the application depth and breadth of data cage in the direction of "information, data, self-flow and integration", establish and improve the supervision and assessment mechanism in accordance with the work requirements of the Municipal Party Committee and the Municipal Government, and promote the work to achieve actual results.[3]

In big data transportation, the matching of precise demand includes several aspects. For example, it is common to predict user demand in advance through big data, so as to formulate corresponding strategies and develop new products, and guide the highlighting and satisfaction of user demand. For example, through the analysis and prediction of existing big data, we can optimize some links in the transportation process, and accurately discover unnecessary and wasted resources. Also for example, we can precisely find the dangers in the transportation process through big data, and precisely push information to traffic participants to avoid unnecessary loss (See Figure 4.3).

In a word, it is one of the most obvious characteristics in the era of big data to accurately mine and meet the demands. In the process of big data transportation, precise demand matching will generate a lot of entrepreneurial opportunities and make future transportation more intelligent.

Figure 4.3 Crane unloading at the dock.

[3]The content is quoted from "http://www.gog.cn/ Guizhou Daily": Big Data + Transportation Management Improves the Efficiency and is Precise and Convenient for the People. The content has been deleted and modified.

4.4 Feature 4 of Big Data Transportation: Real-Time Control

With the gradual improvement of urban traffic, as well as the progress of mobile Internet, big data, and other technologies, people increasingly want to be able to master the traffic information in real time to make choices freely, so as to complete the travel behavior in the most reasonable and effective way.

In the past, real-time traffic information could not be known. Even if there was a traffic jam ahead, drivers could not know. Only when they were stuck in the road could they think of whether other roads were free pass. Such blind traffic mode often wastes the driver's time and traffic resources, resulting in environmental pollution.

With the support of big data and the smart transportation big data platform vigorously built by China and enterprises, people's desire to master the traffic information in real time in the process of travel has been realized.

The intelligent transportation big data visualization platform can support integrated video monitoring system, intelligent checkpoint system, traffic flow detection system, signal control system, and other traffic business systems to realize video monitoring, intelligent checkpoint analysis, traffic operation monitoring, traffic signal monitoring, and other functions, help managers to understand the operation status and change rule of the road network in real time, provide scientific data support for traffic management decision-making, traffic planning and design, and realize remote real-time monitoring of road traffic status. It also realizes the visual management of daily service and duty, helps the managers to master the deployment and dynamics of service personnel in real time to facilitate the timely dispatch and supports the functions of service post distribution, service supervision, service assessment, etc (see Figure 4.4).

Through video monitoring system, detection system, and mobile terminal, the information management of traffic infrastructure is realized, which helps managers to fully perceive and monitor the status of infrastructure, provide support for traffic infrastructure safety management and maintenance management, and improve the level of traffic operation management and service guarantee ability.

At the same time, it can also conduct analysis, summary, integration, and thematic analysis on the historical traffic flow, traffic violations, traffic accidents, and other data to realize the analysis and judgment of traffic

Figure 4.4 Guangzhou Traffic Police Smart Traffic Command Center.

situation, achieve the purpose of scientific and detailed management and provide decision-making basis for the optimization of traffic management departments in terms of traffic organization, police deployment, equipment layout, etc.[4]

Overall, the real-time master of traffic information is shown in the following aspects.

(1) **Real time master for road traffic.** The first need for many people in the driving process is to clearly grasp the traffic conditions of the road ahead. Whether it is traffic jam, traffic accident, road maintenance, or natural disaster; after relevant information is provided to the driver, the driver will choose his own route according to experience and the recommendation of map software.
(2) **Real-time master for public transportation at all times.** The pace of development of the whole society is moving faster and faster, and people have increasingly higher requirements for the accuracy of time. When choosing public transportation, it is very important for

[4]The content is quoted from Digital Hail (WeChat), "Big Data Visualization of Intelligent Transportation Makes Urban Operating Vehicles be Monitored and Controlled".

people to accurately know the schedule of buses, metros, and other means of transportation.

(3) **Real-time master for traffic safety.** The most obvious performance is that when relatives travel, whether by train or by plane, people want to know the real-time situation of the relatives' travel. If their relatives know when the train and plane are running in real time, they will also worry less. For example, in the APP of "Flight Manager", there are real-time flight information, such as flight take-off, flight status in the air, landing and other information can be seen at a glance.

(4) Government agencies can control traffic conditions in real time to take better measures to guide traffic, such as traffic lights control and temporary traffic control.

(5) Enterprises can control traffic information in real time and provide better services for the public. For example, Baidu map, Amap and other map software, when providing navigation services for travelers, must master the traffic conditions in real time, otherwise it may mislead the travels (see Figure 4.5).

4.5 Feature 5 of Big Data Transportation: Efficiency and Convenience

The advantages of big data transportation are clear: it helps transportation managers to efficiently manage all elements of the transportation field, and it also helps to access transportation facilities for transportation participants conveniently. Currently, the government and enterprises are

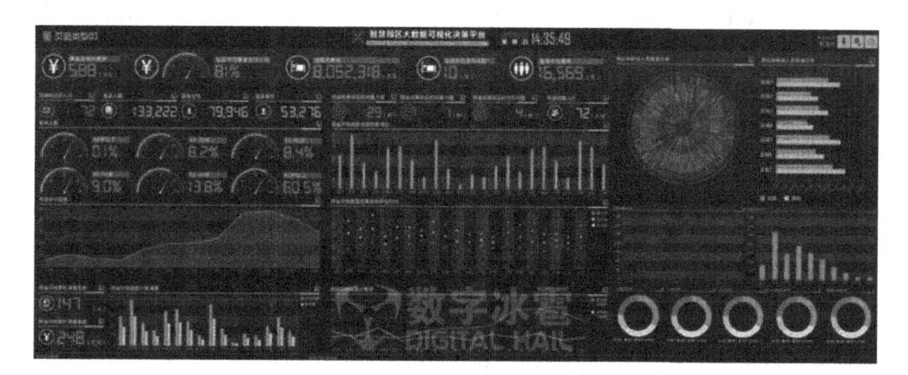

Figure 4.5 Big data visualization decision-making platform.

Source: Digital Hail.

working on relevant big data transportation platforms. These platforms add a lot of convenience to the public while ensuring the efficient operation of transportation.

For example, the "Rail Transit Big Data Expert System" on display at the 2017 Beijing International Urban Rail Transit Exhibition (UrTran) is a microcosm of the efficiency and convenience of big data transportation.

The system is built on the Fiberhome FitData big data platform, using big data analysis and vehicle healthiness models for vehicle equipment operation health assessment, intelligent fault diagnosis and operational risk early warning, to achieve intelligent operation and maintenance of vehicles and equipment. It can effectively improve the locomotive and equipment health operation management level of rail transit operation enterprises and provide expert intelligent operation and maintenance services for locomotive and related equipment safety operation of rail transit operation enterprises.

The expert knowledge base is one of the features of the expert system of rail transportation big data.

(1) Fault analysis is based on existing fault data, with big data analysis showing the comparison number of different faults and also calculating its failure correlation.
(2) Different fault structure trees are used for different faults, which is more convenient for fault analysis and location. The expert system supports to constantly modify the fault tree structure with practical experience data and machine learning technology, add, modify, or delete events in nodes and constantly improve the fault tree.

Through the use of the system, it can effectively improve the reliability of equipment operation, improve the efficiency of operation and maintenance and reduce operation and maintenance costs. According to the clustering analysis of operation and maintenance data, vehicle equipment manufacturers are provided with improvement proposals for design and manufacturing.[5]

Such big data transportation platforms and systems, in addition to providing the rail transit operators with corresponding information,

[5]The content is quoted from Rail Transit World, "Big Data and Artificial Intelligence Make Rail Transit Operation and Maintenance More Efficient and Convenient". The content has been deleted and modified.

solutions, and improvement suggestions, more importantly, they ensure the normal operation of rail transit and ensure that people can travel in a timely and accurate process. Otherwise, in the first-tier cities like Beijing, Shanghai, and Guangzhou, once the rail transit breaks down, it will affect the normal work and life of huge number of travelers. If the work and life of travelers are affected, the economic development of the whole society will be damaged. Therefore, the efficiency of big data transportation is related to many aspects of the country and the people.

4.6 Feature 6 of Big Data: Intelligence

The big data transportation and intelligence are closely linked. When big data transportation becomes more efficient, convenient, self-service, and free, people's travel will become more intelligent.

When waking up in the morning, your mobile phone or the Internet of Vehicles will push the real-time traffic conditions on your way to work, so you can go out earlier or take other detours depending on the actual situation.

When traffic congestion is encountered, the detection equipment at the roadside can detect and alarm, and traffic management departments will choose the best solution based on data analysis to ensure the rapid resolution of traffic congestion.

In the event of a traffic accident, the Internet of Vehicles system can automatically call for assistance to protect the safety of the travelers.

The so-called intelligence is through the analysis and mining of big data, traffic participation in all links and the information transmission can be autonomous and self-help, the traffic managers can respond in a timely manner and all traffic behavior can be run in accordance with the established rules in an orderly manner. The efficiency is greatly improved and the safety index is greatly enhanced. And in all of these, big data transportation has been perfectly addressed in the development process.

For example, in order to obtain traffic data in time and accurately and build a traffic data processing model to lay a good foundation for future construction of intelligent transportation, Jinan municipal government, through big data technology, controls 4,764 lanes in real time in the city, 196 signal lights in the intersection, realizing the regional adaptive signal control. If there is road congestion, it will automatically alarm and the traffic flow will be automatically detected.

(1) Relying on the "cloud" on the police affairs to control 4,764 lanes in real time. According to the idea of "dynamic defense, precise command, integrated operation and overall prevention and control", relying on the police affairs cloud big data processing technology, four lines of defense constructed by 1,758 points are built, 4,764 lanes are under real-time control, real-time data analysis mode and distributed storage architecture are adopted to realize the real-time processing of traffic data at any checkpoint in the city, real-time display of vehicle operation track, accurate query and comparison of historical traffic data, the playback of driving track on GIS map, the interconnection and interworking of various police type resources between platforms, so that remarkable results have been achieved in actual combat.

In addition, Ji'nan has also developed the public transport priority signal control system and the traffic management application systems such as taking photos of occupied yellow grid line and uncourteous pedestrians. Twenty-two priority signal control points have been built in the urban area and 255 sets of public transport onboard snapshot and illegal snapshot in road section have been installed. The annual average number of illegally occupied public transport lanes to drive is more than 150,000. Thirty-nine sets of yellow grid line occupation snapshot equipment and 5 sets of unyield to pedestrian snapshot equipment have been installed. There were 7,574 traffic violations of occupied yellow grid line and 24,136 events of unyield to pedestrians by snapshot. The occupation of public transport lanes, non-motor vehicle lanes and the scramble for traffic are significantly reduced, the civilized driving behavior of courtesy zebra crossing is gradually increasing, and the space for public transport and slow traffic is effectively guaranteed.

(2) Prevention and control of traffic jam: If there is the traffic jam, it will alarm automatically. Ji'nan actively promotes the construction of intelligent video monitoring system, improves the function of 612 analog videos and increases the functions of automatic alarm for abnormal events such as road congestion and automatic detection of traffic flow. At the same time, Ji'nan has deepened the integrated application of high-definition digital video monitoring, built 481 high-definition video monitoring points, and 368 multi-functional electric alarms, and they have greatly improved the image quality, which realizes the automatic collection of all kinds of dynamic and static traffic violations, supports the application expansion of the secondary image recognition technology, forms the organic combination of the city's low point and high point monitoring and provides real-time video signal support for daily traffic management and emergency scene disposal.

On the other hand, Ji'nan has deepened and upgraded the upgrading project of signal control system, built a convenient and efficient accurate traffic signal control system and regional adaptive control system, updated the old equipment, and upgraded the existing traffic signal system, which truly realizes the synchronous adjustment of signal timing with the change of traffic flow, establishes and improves the signal remote control system based on the command center at all levels, realizes accurate, reliable and convenient signal coordination control, reduces the delay of driving and parking and improves the traffic efficiency.

(3) The official account of Ji'nan traffic police has launched three modules and eight functions. Under the mode of "Internet +", intelligent transportation is inseparable from the value orientation of facing the people's livelihood and expanding traffic management services. The ultimate effect of "Internet +" is to go deep into thousands of households, eliminate the links restricting innovation in the past, connect the isolated innovation and let information serve the public.

Taking mobile Internet technology as the carrier, Ji'nan public security has built a public service platform for traffic police in Ji'nan, which covers multiple information channels such as APP, WeChat, and Weibo to directly reach the mobile phones of the public to carry out traffic management business and various travel related services, and the whole people participate in jointly improving the quality of urban life, which has been listed as one of the 18 "practical things for the people" by the Municipal Party Committee and municipal government this year.

WeChat service account has launched three modules and eight functions of rapid accident handling, traffic management business, and account binding and authentication, providing services such as rapid accident handling, vehicle snapshot and query and driver license query for the public, with a total of 115,000 followers and 31,000 certified drivers and 21,000 bound vehicles. The public service APP is constructed in two phases, and the APP is under internal test. In the first phase, multiple modules, such as personal center, traffic management business and Information Center, are launched to provide more than 20 functions for the public, such as notification and push, travel conditions, traffic management information, business query and handling, peripheral services, opinions and suggestions. In the second phase, more than 50 functions will be realized on the basis of the first phase, so as to expand the field of convenient and beneficial services.[6]

Through this series of measures, Ji'nan has made the city transportation more intelligent, in which big data plays an important role. In the future, with the accumulation of transportation data, any imperfections in the transportation link can be manifested through data. The big data can provide solutions and this will help the transportation to become more and more intelligent.

[6]The content is quoted from China Security Exhibition Network, "Under Big Data, Intelligent Transportation is More Convenient in Ji'nan and Ji'nan Benefits a Lot". The content has been deleted and modified.

Part II

Application: Practical Implementation of Big Data Transportation

Part II of the book discusses the implementation of the big data transportation, that is, how to import and apply the big data transportation in various fields will be illustrated by cases including competitions of big enterprises in the field of the big data transportation, big data to help public transportation planning, big data transportation and logistics, the future of big data transportation, etc.

Chapter 5

The Entrance of Big Data Transportation

In recent years, "entrance" has become a popular word and some people say that if you find the entrance, you will find the way out. More enticing to say, those who have entrance can obtain the world. For big data transportation, the so-called entrance is the most basic value point. Through the scientific creation, management, and operation of the entrance, the value points on the big data transportation platform can be connected with each other, and a data value chain can be derived so as to better promote the efficiency of the big data transportation system.

5.1 T-Union

In November 2015, the 13th Five-Year Plan proposed to promote low-carbon development of transportation and gave priority to public transportation. Some cities have introduced the intelligent public transportation system, and the public transportation IC card charging system (T-union) is the most widely used subsystem in the system.

The total number of T-union (Ling Nan Tong) issued in Guangdong Province has reached 46 million, covering the public transportation fields of public transport, subway, and taxi. In terms of geographical scope, 21 cities including Guangzhou, Shenzhen, Dongguan, Foshan, Zhaoqing, Jiangmen, and Shanwei have been interconnected, and the perfect connection with Hong Kong's Octopus and Macao's Macau Pass has been achieved. Ling Nan Tong plans and collects the data of the whole province in a unified way. The maximum daily data volume is more than 12 million, and each data contain at least 20 attributes. It includes the cardholder's transaction time, place, transaction type, transaction frequency, card issuance data, customer service data, recharge and consumption data, terminal record data, bus stop information, line information, etc.[1]

At present, the T-union issued by the public transportation department is widely used, resulting in a large number of swiping transactions in operation every day. The transportation department can accurately collect the travel information recorded by it. As a result, the application field of the T-union system continues to expand, which carries a large amount of traffic data, and naturally constitutes an important entrance to the big data transportation system.

5.1.1 *Accuracy of the entrance of T-union data*

The maturity and universality of T-union system can ensure the accuracy of data sources. In Beijing, a special management system has been set up in the transportation management organization, which can accurately count the flow of people in the metro within a day, or extract the travel route of a group of users. In addition, the traffic volume of people who use the T-union to take the bus every day can also be accurately counted. Even if some users choose to use cash to take the bus or metro, the statistics department can also calculate the number of people in a day according to the proportion.

In the long run, the one-day travel of urban residents usually has regression characteristics. The specific performance is from the place of residence to the end of residence, so that the bus stops have a special symmetrical effect. Compared with other data sources, the multi-day travel data of residents recorded by the bus T-union information, even though occasionally changed, is basically the same for a long period of time, which also ensures that the data collected by the bus T-union can accurately determine the travel destination of passengers.

This kind of precise collection effect, does not need to carry on the massive analysis and judgment to the passenger's next station point in the data follow-up analysis, thus effectively reducing the complexity, and improving the efficiency and the precision.

[1]It is quoted by Yu Hongling, Liang Liang, and Wu Jiayin, "Construction of Big Data Application System of Urban T-Union". The content has been deleted and modified.

5.1.2 *Scientific features of the entrance of T-union*

Basic data of urban public transportation can be divided into static data and dynamic data. The dynamic data are mainly updated in real time with time, including line passenger flow, station passenger flow, traffic flow, real-time vehicle speed, and other data. The static data refer to the data that will not change or need to be updated in real time within a certain period of time, such as bus station location data, station information of each bus line passing, distance between stations, and location information of transfer stations.

Through collecting and providing static and dynamic data at the same time, T-union ensures the scientific features and comprehensiveness of the data obtained and builds a good foundation for further data retention and analysis.

5.1.3 *Wide application of T-union data*

With the support of a wide range of operating platforms and huge data, the collection and application of T-union data present a diversified development status. The acquisition and analysis of passenger flow information and operation information are becoming more timely, accurate, and wide.

For example, due to the wide coverage of the T-union system and the strong instantaneity of data, its data results can be widely used. Through the collection of relevant data attributes, the public transportation department can calculate the travel parameters which can be divided into time periods, road sections, people to study, and apply and even create a public transportation model.

In addition, using the data collected by the T-union, the public transportation management department can take targeted measures to formulate response plans in different situations in advance. For example, scientific distribution of transportation capacity, timely handling of various traffic problems and the improvement of the pertinence of decision-making are typical big data applications.

5.1.4 *Future development of T-union data collection*

Using T-union information to collect and mine data has the advantages of large amount of data, easy processing, and strong real time, which can save a lot of costs for the construction of big data transportation system.

At the same time, there is a considerable space for future development in the data collection of T-union.

So far, NFC near-field payment function of mobile phone has not been popularized in China's consumer groups. Accordingly, data collection of T-union has not been combined with mobile payment. With the development of science and technology, once the NFC payment function is popularized, the ability and effect of data collection of T-union will be improved together with the data collection of mobile payment system. At that time, the T-union will become a more important data entrance, which will more accurately output the travel scale of urban residents, relevant information, and other data for the transportation management department (see Figure 5.1).

Further, at present, the national intelligent bus card does not implement the joint system like the bank card; the ticket price in many cities is only related to the length of the route, and the public transport in some cities is still very uncertain. Therefore, with the addition of new elements of urban development in the future, further improvement of the performance of the T-union system, such as improvement of the deconstruction of public transport IC card database, establishment of public transport data warehouse, integration of public transport basic information, IC card information, and public transport enterprise operation management information, and the realization of information fusion can make it play an important role in the realization of data entrance in the era of big data.

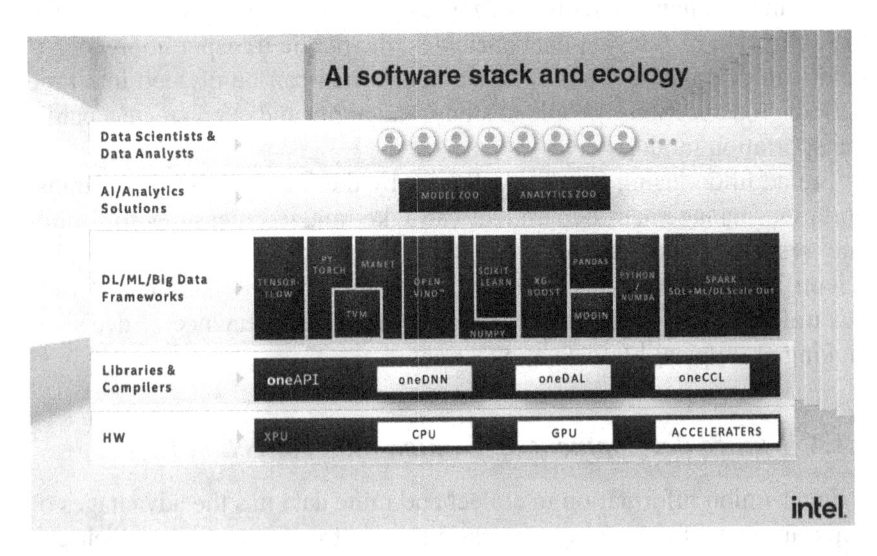

Figure 5.1 AI software stack and ecology.

5.2 GPS and BeiDou System

The U.S. GPS satellite positioning system was designed and developed by the U.S. Department of Defense in 1973 and completed in 1993. Its space part consists of 24 working satellites. The distribution design of these satellites makes it possible to observe more than four satellites anywhere and at any time in the world. The location of the receiver can be determined by the solution of the positioning information.

In 1994, China began to build a "BeiDou First Generation" satellite navigation and positioning system. It is expected that by 2020, a global "Beidou-2" satellite navigation and positioning system composed of more than 30 satellites that can cover the whole world will be built. Its construction goal is to provide land, sea, and air navigation and positioning services for our military and civilian users in China and surrounding areas.

At present, both GPS and BeiDou system carry the functions of vehicle tracking, emergency assistance, information query, travel planning, etc. In addition, in order to analyze the traffic situation with big data, the real-time information collected by GPS and BeiDou system is used to integrate other traffic data and form data sources, which also makes these two systems become the important and high-quality data entrances.

In 2016, the traffic data studio of Wuhan Traffic Management Bureau built a big data traffic guidance system based on the Internet. In the column of real-time data source access on the large screen of this institute, 260 million GPS data have been achieved. Through these data, it can show the thermal conditions of the city's roads from 0:00 to 24:00. It can not only reflect the current congestion location in real time but also predict the congestion section within one hour.

Before using GPS data source, once the road is congested, it can only be handled on the spot after confirmation by means of public alarm and police video inspection. At present, the big data management system based on GPS can alarm immediately in case of congestion. Through the analysis of GPS data of the whole city's roads, the traffic management system can better command and dispatch decisions, intervene manually in advance and effectively alleviate congestion.

Similarly, BeiDou satellite positioning system is also providing active and sufficient data sources for big data transportation.

In April 2017, all buses in Pudong, Shanghai were equipped with BeiDou satellite positioning chips, which sent signals to the background dispatching system every 10 seconds. Compared with GPS,

BeiDou satellite positioning chip uses "short message" technology, which can transmit data in poor communication environment. Using the real-time data, the public transportation management department constructs the intelligent cluster scheduling system. With this powerful background support, the vehicle running track can be real-time formed on the station board through the receiver installed on the bus station pole, so as to provide the arrival forecast for passengers.

No matter GPS or BeiDou system, they both have the characteristics of accurate positioning, convenient terminal, rapid installation, and convenient data extraction. Therefore, GPS and BeiDou system can record the dynamic changes of urban traffic and crowd movement at any time, so as to provide a lot of important data information for the research of urban data transportation management. Compared with the traditional methods such as questionnaire and meter, the data information is more abundant and accurate and it can be used directly.

5.2.1 *Data application of taxi onboard system*

Taxi is an indispensable part of the urban transportation whose speed and traffic density can largely reflect the road traffic conditions. Therefore, taxi trajectory data reflected by onboard GPS and BeiDou system can be used to analyze the changes of urban traffic congestion. For example, traffic flow density model is built and large taxi trajectory database is used to automatically determine the capacity of each road section, so as to estimate the average speed of the road section and predict the traffic situation and potential traffic congestion.

Because the changes of urban traffic flow usually follow a specific pattern, taxi drivers usually choose a relatively fixed route. By using GPS and BeiDou system to mine the long-term running data of taxis, not only can the normal trajectory be identified automatically, but also the abnormal trajectory and accidents such as traffic accidents and main road closure can be found and managed in time.

5.2.2 *Operation and management and support of urban traffic*

Through further statistical analysis of the data provided by onboard GPS and BeiDou, different indexes and data can be obtained, such as driving mileage, passenger carrying journey, space-time rate and spatial-temporal characteristics of distribution. These information can be used not only for daily management of taxi and bus, but also for public transportation management department, and are as the basis of decision-making for taxi or bus.

Further thinking, if the data accumulated by the aforementioned onboard equipment is mined, the rules hidden in the data can be further discovered so as to provide guidance and decision-making basis for urban traffic planning and management and optimize urban traffic.

5.2.3 *Movement data application of smartphone*

In recent years, the use of smartphones is becoming more and more popular. At the same time, the GPS or BeiDou system equipped with smartphones can also be used as the traffic detector of the target point to collect the real-time data of the fixed point for predicting the traffic flow speed.

Traditional traffic flow data collection relies on ground induction coil to obtain data, but the coverage of ground induction coil is small, the cost is high and the popularity is low. In contrast, the transportation management department can cooperate with the wireless operators to obtain the call data, the SMS data, the mobile phone handover data between base stations, the location data of the base station where the mobile phone is located, as well as the change of mobile phone information volume, and can obtain the road section traffic data in time by using the mobile phone location system.

After collecting road information by mobile phone, relevant organizations of the transportation management department can estimate and predict the corresponding traffic conditions including travel time and congestion through training statistical model so as to provide convenient travel planning services for mobile phone holders and even all travel vehicles.

What is more, the traffic data collected by mobile GPS or BeiDou system can show the specific movement mode. Movement mode is an important attribute to describe people's traffic behavior. Generally, the movement mode includes the driving, cycling, bus, metro, and walking. After introducing the concept of movement mode, the data of GPS or BeiDou system can be further divided into different movement modes, and then reasonable management or suggestions can be made according to the mining results.

Through any device that provides GPS or BeiDou functions, it can collect massive traffic data. But in most cases, the transportation management department cannot directly use these data. This is because the amount of relevant data is very large, and a single point of GPS or BeiDou system has not much significance for the application of traffic management in the whole city. In the future, relevant research and management institutions must mine frequent cycle patterns from the data of GPS and BeiDou system, and find out the movement and living rules of residents in different regions and communities, as well as on different roads in a city through analyzing cycle patterns. At the same time, the cycle mode can also be regarded as the compression of the traffic movement track. The data of GPS and BeiDou system can be cyclized and the space of storage and processing can also be saved.

5.3 Internet of Vehicles

The concept of Internet of Vehicles is proposed based on the Internet of Things in China. In September 2009, the Fourth National GPS Operator Conference was held in Shenzhen, where "Internet of Vehicles" appeared for the first time. In November of the same year, the definition of the "Internet of Vehicles" was formally formed at the Fourth International Symposium on RFID Technology Development held in Shanghai.

The Internet of Vehicles refers to the electronic tag loaded on the vehicle. Through the identification technology such as radio frequency, now all the attribute information, static, and dynamic information of the vehicle are extracted and effectively used on the information network platform. According to different functional requirements, the Internet of Vehicles will be able to support the operation status of all vehicles, effectively supervise and provide comprehensive services.

Of course, the aforementioned definition is mainly made from the perspective of information technology. With the development of technology and society, the role of the Internet of Vehicles cannot be fully reflected in it. In essence, the significance and value of the Internet of Vehicles lies in the collection and processing of data and information of each vehicle in the urban traffic network, the sharing of data, the common connection among vehicles, roads, and the Internet so as to establish a traffic management system independent of vision, weather conditions, and human operation.

The following is a typical case of using the Internet of Vehicles to obtain and apply data:

On-board automatic diagnosis system (OBD) was first used in the United States to detect different types of vehicle exhaust emissions. Through this interface, we can read various data of vehicle operation, including engine data, driving behavior and fuel consumption data, vehicle safety data, and so on.

In China, a start-up company has launched small hardware that can be plugged into the OBD interface of the vehicle. The hardware transmits the useful data in the car to the mobile phone through Bluetooth, and then to the big data platform through the mobile Internet. Through this channel, the data in the vehicle can be extracted and transmitted, and the fuel consumption, driving behavior and track can be recorded, so as to remind about the vehicle safety information.

In fact, the data source of the Internet of Vehicles is extensive, including the communication among the in-vehicle, the vehicle and road, the vehicle and vehicle, the vehicle and the outside world, the vehicle and the people. If we collect and excavate it comprehensively and carefully, we can achieve the degree of harmony and unity of "people, vehicle, road and environment".

5.3.1 *Data exchange between vehicle and road facilities*

Through the construction of the Internet of Vehicles, data exchange is compatible with the data transmission between traditional vehicles and traffic signals (such as signs, driving lines, speed limits, traffic lights, etc.). After that, through obtaining the query information and navigation request of the networking vehicles, the required data is returned to the related networking vehicles. In this closed-loop system, the vehicle connected to the Internet of Vehicles can obtain the digital information of the traffic signal system through the wireless data transmission module in the vehicle, and display it directly in the vehicle and connect the information with the driving system in the vehicle as the reference signal of the driver.

5.3.2 *Interconnected and interactive data of vehicles*

Through the installation of each wireless data transmission module in the vehicle, different vehicles can send traditional traffic indication information and digital information at the same time. In this way, data synchronization can be formed between related vehicles, digital signals of different vehicles can be obtained and displayed in the vehicles at the same time and the information can be connected and interacted with the driving system in the vehicles, which becomes the basis for safe driving of vehicles.

5.3.3 *Data exchange between vehicle and transportation management department*

The Internet of Vehicles system is divided into three parts: onboard terminal, cloud computing processing platform and data analysis platform. According to the requirements of traffic management industry, the Internet of Vehicles can effectively monitor and manage the vehicle data.

In the actual traffic, the specific vehicle operation often involves many data, such as switching value, sensor analog value, and CAN signal data. When the driver is operating his or her vehicle, the generated vehicle data will be sent back to the background database to form massive data, and the cloud computing platform will realize the "filtering and screening" of massive data, and then the data analysis platform will process the data in a report form so that the managers can view it directly.

5.3.4 *Data integrated service of Internet of Vehicles*

By acquiring real-time data through the Internet of Vehicles, an accurate and efficient comprehensive traffic management and control system can be established to effectively connect "people, vehicles, roads and environment". Besides, according to different functional requirements, the traffic management department will conduct effective digital supervision on the operation status of all vehicles and provide comprehensive services.

For example, data services such as collision warning, electronic road sign, traffic light warning, online vehicle diagnosis, and road slippery detection can be used to provide drivers with immediate warning and improve driving safety; travel efficiency of residents can be improved through traffic congestion detection, path planning, road charging, etc.

With the development of big data transportation management, the Internet of Vehicles as the data entrance will create a new management way. At this stage, the transportation management department must be guided to establish an effective coordination mechanism and a complete industrial chain among vehicle manufacturers, IT enterprises, and consumers and form industry-related technical standards so as to promote the growth of the Internet of Vehicles industry.

5.4 Road Network Monitoring

The road network monitoring whose full name is the highway network monitoring system takes the highway network monitoring center as the command headquarters, which is responsible for coordinating the traffic monitoring and control of the whole road network and for the unified coordination, command, scheduling, data collection and analysis, and statistical report of the traffic monitoring system of the whole road network. Through the scientific application of the road network monitoring system, the traffic management department can obtain various data in time, and make statistics and analysis on the road condition information and road service level.

During the Spring Festival in 2017, the real-time road conditions and trend judgments of the national road network were continuously rolling on the various media such as CCTV's News Live Room, China Traffic Broadcasting, the Ministry of Transport, and the WeChat official account. This operation research and analysis report on national road network, during the Spring Festival, strengthens the real-time detection, emergency response, and information release of the road network operation, reflecting the important role of the road network monitoring system.

In 2016, the road network center of the Ministry of Transport strengthened the cooperation with different social resources, together with the units such as the Planning and Research Institute of the Ministry of Transport, the Research Institute of Highway Ministry of Transport, China Transinfo, Chinese Academy of Sciences and Amap and successively developed the highway condition operation information management business system (GIS-T comprehensive business system), the national highway network real-time road condition system, the national highway congestion and flow trend system and the comprehensive road condition business system of Beijing–Shanghai expressway, which were put into trial operation during the Spring Festival in 2017.

On January 28, through the national highway network real-time road condition system, the road network monitoring personnel on duty found that there was congestion in several sections of the Ya'an–Xichang section of Sichuan Province on the Beijing–Kunming expressway, with the length of the congestion reaching dozens of kilometers and the speed below

20 kilometers per hour. The personnel on duty promptly contacted Sichuan Provincial Department of Transportation for verification and understanding. After finding the causes, the local road administration personnel and expressway traffic police took remote counting, fixed-point diversion, multi-channel information release and other ways to strengthen the guidance. By the afternoon of January 31, the congestion in Ya'an–Xichang section was relieved.

Through the application of the road network monitoring system, it can timely and accurately collect all kinds of data information of road traffic, calculate, and analyze the data information so as to make scientific decisions, and then realize the management intention quickly by optimizing control. In addition, the road network monitoring system can also support the release of effective information by collecting road data information, provide comprehensive information services for drivers, then guide drivers to drive correctly and strive to maintain the dynamic balance of urban traffic.

The road network traffic monitoring system mainly realizes the collection and application of traffic data through the following ways:

(1) **Traffic Flow Detection System:** Many cities have developed and used traffic flow detection systems, especially for urban traffic conditions and mixed traffic flow characteristics. The similar system adopts the corresponding intelligent digital image information processing technology, which has many characteristics such as advanced technology, stable performance, no damage to the road surface, strong applicability and convenient installation and operation.

The real-time response software is used in the traffic flow detection system. The system sets the vehicle image information collection camera on the top of the lane to be detected. The video-trigger recognition software is used to analyze the digital video signal sent back by the camera to determine whether there is a vehicle passing by. When a vehicle enters the detection area, the traffic flow detection module is immediately started, and the system will automatically add one for each vehicle passing by.

In a very short time, the traffic flow detection system can complete the traffic flow detection, queue length, waiting time, lane occupancy, road information analysis, and retrieval data processing. These real-time data can be compared with the traffic flow parameters in the database, judge the road congestion and carry out remote data release, update, query and other processing.

In addition, the system can also count the passing vehicles, including the traffic flow, waiting time, lane occupancy, and other data according to the date of. The data results can be the output and printed in the form of tables, histograms, and graphs, and transmitted to the traffic management center on time. The management department can take corresponding measures to disperse the traffic according to the data to meet the requirements of intelligent management and monitoring of urban traffic.

(2) **Road Condition Detection System:** The system is mainly used to collect the original data on the road, mainly relying on vehicle detector, detection coil, various communication equipment, etc. These machines are set up in the main position of urban roads, and then installed with cameras and video equipment so as to obtain the actual data of road operation.

Moreover, the emergency signals on the road can also be obtained through the roadside emergency telephone. Through the complete meteorological information collection system, the ground humidity, temperature, visibility, wind direction, wind speed, rain and snow days, and other different meteorological data can be detected so as to provide the precondition for the reasonable distribution of the road operation support capacity. At the same time, the aforementioned data can also be displayed in a centralized or individual way through TV wall, computer, and Internet, which can provide reference for vehicle owners' driving and traffic management.

Further, in order to better realize the data collection function of road network monitoring, it is necessary to guarantee the ability of data acquisition of outfield monitoring, that is, to avoid blind spots of monitoring. At present, the outfield monitoring is divided into two modes: automatic mode and manual mode. Road environment data, traffic data, meteorological data, and abnormal data are adopted, respectively.

5.4.1 *Technical advantages of road network monitoring*

In terms of data source entrance, road network monitoring has its unique advantages. First of all, road network monitoring can achieve multi parameter and large-scale detection. On the one hand, network monitoring can quickly extract a variety of traffic parameters in a large area from video signals, and extract parameters that are difficult to detect by other methods. On the other hand, the road network monitoring system can cover 4–6 lanes with one video signal at the same time and realize traffic flow detection in a large area.

Secondly, road network monitoring can achieve multi-stage detection, that is to say, it can detect the traffic demand and related variables of the intersection, and also detect the specific data of the implementation of the signal control scheme so as to complete the feedback of the whole process effect of the automatic signal control.

In the installation and use, the road network monitoring system can complete the reset of the detector without damaging the road surface or interfering with the normal traffic. Compared with the terminal installed on the vehicle, the use method is easier. More importantly, on the basis of the system, more practical methods for data acquisition can be developed.

In a word, a perfect road network traffic monitoring system is a mature data entrance of big data transportation system whose correct use can improve the efficiency of road traffic management, give full play to the value of road use, ensure traffic management and that the traffic system is operating smoothly.

5.5 Electronic Navigation Map

If big data transportation management system is changing everyone's travel and life style, the electronic navigation plays an important role in the big data transportation system, which directly affects the success or failure of big data transportation system.

In recent years, due to the continuous expansion of the scale of China's navigation and location service industry, the electronic navigation map industry has developed rapidly. In 2010, the market scale of China's electronic navigation map was about 1.128 billion yuan, an increase of 35.17% over the previous year; in 2015, the market scale of China's electronic navigation map climbed up to 3.691 billion yuan. Among them, the scale of onboard electronic navigation map industry continues to expand and maintain a high growth rate. In 2015, the market scale of onboard electronic navigation map reached 2.057 billion yuan, with an annual increase of 17.61%.

In 2016, a large number of taxis in Wuhan were equipped with a new onboard electronic navigation map system. When using the electronic navigation map, the driver and the passenger only need to set the starting point and the ending point in the system, then they can choose the driving route freely. For the traffic management organization, the electronic navigation system has a strong data collection ability, with information screen, service evaluator, monitoring equipment, and intelligent roof lights to ensure that all taxi information in the city is synchronized. The management department can carry out intelligent network management of taxi and driver through GPS positioning, operation, service evaluation, cost, industry certification and other information, greatly improving the efficiency.[2]

In essence, electronic navigation map is a kind of spatial data system based on computer system and it has the characteristics of rich and diverse data content, complex data relationship, and information sharing. At the same time, electronic navigation map itself is a kind of geographic data system, which has most functions of geographic data system. Therefore, electronic navigation map can display digital signal (including digital map, remote sensing digital image and collected digital data) and analog signal on computer screen through Internet.

In fact, due to the combination of the visualization function, data query and analysis function, multimedia technology and virtual reality technology of digital mapping technology, the electronic navigation map has won the widespread interest of traffic management departments since its emergence. With GPS navigation, BeiDou navigation, and other applications, the electronic navigation map is constantly providing data information for big data transportation management system.

The electronic navigation map needs a large amount of data to be collected, which must be obtained by a variety of comprehensive ways to ensure that the information is comprehensive, real time, and reliable. The main ways to obtain data information are as follows:

(1) **Direct Data Obtaining and Sharing:** For most of the basic geographic information, countries, most regions, and cities have established their own basic geographic data information systems. For example, China has established a basic geographic database with a scale of 1:10000 and 1:2500, and most regions and cities have

[2]The content is quoted by *Changjiang Daily* on January 22, 2016, Fu Sha, Reporter, and Ke Xuan, Correspondent.

established a basic geographic information database with a scale of 1:500 and 1:200.

These basic data systems, as an important carrier of traffic information, have played an important role. For these existing data information, they can be directly used as the data source of the electronic navigation map. After being organized, they can be directly input into the electronic map database of the traffic management department.

At the same time, the traffic management department can also share the data of other systems into the electronic navigation system. For example, traffic light control system, bus scheduling dynamic change table, and vehicle tolling system contain a lot of traffic map special data. If these resources are integrated, the traffic map data collection will be better completed under the condition of ensuring the accuracy and comprehensiveness of the data.

(2) **Data Acquisition with GPS and BeiDou System:** Through the combination of electronic navigation map, GPS, and BeiDou system, the spatial data information of point elements can be accurately collected, such as vehicle location and traffic accident location.

Because GPS and BeiDou system can provide global coverage, all-day free high-precision navigation and positioning services, the onboard receiver can match the information obtained by the system with the electronic navigation map. In this way, the specific location data of vehicles in the traffic road can be submitted to the traffic management department.

(3) **Data Sharing with GIS Intelligent Traffic Supervision System:** By connecting the electronic navigation map with the GIS intelligent traffic supervision system, it can directly help the traffic supervision organization to realize the following data operations on the map:
    (a) Stepless enlargement and reduction of electronic navigation map;
    (b) Translation and switching of electronic navigation map;
    (c) Display mode optional;
    (d) Arbitrary call of dozens of layers;
    (e) Graphic editing;
    (f) Query of various information (geographic target, vehicle information, etc.).

For example, in 2009, the GIS system of Xi'an Traffic Command Center integrated the electronic navigation map system so it has many map

Figure 5.2 In-taxi surveillance images.

Source: Wuhan Transportation Network.

data operation functions, such as map roaming, map zooming, query of site facilities in the map, and "eagle eye" rapid positioning (see Figure 5.2).

Through the use of map data, the system can display the dynamic and static data of video equipment, signal machine, guidance screen, onboard GPS monitoring terminal equipment, electronic police, and other dynamic and static equipment, at the same time, it can display them in the GIS map display screen. We can just click on the map and integrate the detailed traffic dynamics and it can be immediately displayed on the computer screen.

Electronic navigation map is an important entrance of big data transportation management. Traffic management departments must coordinate enterprises, society, researchers, and users in an all-round way to form a comprehensive force to ensure that electronic navigation map will give full play to the value of data collection and mining in the future.

5.6 ETC

In recent years, electronic toll collection (ETC) technology has become more and more mature, and large-scale industrial applications have been fully developed. On May 1, 2007, the national standards for the ETC of intelligent transportation system were officially promulgated and implemented, and the construction and promotion of the ETC system was carried out again on the basis of the whole country. In May 2009, Beijing ERC electronic toll collection system began to operate.

In the process of daily toll collection, the ETC system records the original data of vehicle traffic which contains a lot of traffic flow information. Therefore, the ETC system has the function of providing statistical data and information services for big data transportation system.

Liu Erwei, who has worked in Beijing for many years, got the car number at the beginning of the year. He listened to his friends that it was very convenient to install ETC to go back to his hometown in Henan at high speed. Therefore, when he bought a new car, he installed the ETC.

Recently, Liu Erwei drove back to his hometown with his family. Liu Erwei said, "from Beijing to Luoyang, we passed four or five toll stations without waiting in line once. Compared with the manual toll gate with the long line, we are like putting in wings".

According to the data from the Ministry of Transport, there are 50 million users who have installed the ETC like Liu Erwei in China. What does 50 million mean? By the end of 2016, China's car ownership had reached 194 million, and the total number of private cars reached 146 million, which means that an average of one vehicle in four had installed the ETC.

By the end of April this year, the National Expressway ETC users have covered 29 provinces, the ETC Lane coverage rate of the main toll station is more than 98%, and the daily traffic volume accounts for more than 30% of the expressway traffic volume.[3]

At present, the foreign ETC system is relatively perfect, but there is a big gap in the penetration rate of the system, collection of charging information and service in China. How to make use of the ETC toll data to build the data entrance of expressway traffic operation under the condition of low popularity and less collection means has important research and practical significance.

---

[3]By *Economic Daily* on May 24, 2017, Reporter Qi Hui.

5.6.1 *Highly efficient collection of vehicle passing information*

Through the ETC system, the personal information of the car owner, the amount in the card and the data information such as the speed, time and path can be obtained efficiently. In terms of data acquisition, the ETC is far superior to the traditional vehicle information collection methods such as camera monitoring, license plate recognition, and ground induction coil. The traffic information collected is more comprehensive and accurate.

In addition, in important areas such as the entrances and exits of customs, ports or scenic areas, vehicles that use the ETC can directly enter without parking, providing basic data for the management department. For the road sections with high data accuracy and quality requirements, through the analysis of daily vehicle access records, it can also check whether there is abnormal traffic operation through data.

5.6.2 *Provision of analysis report to competent department*

The ETC system in the expressway can record the car owner's information and charge amount, while the ETC system in the city can record the daily driving time of the owner. Making full use of these data can maximize the value of ETC.

In fact, at present, no matter the expressway ETC network or the single ETC in the city, there are mature technology applications in data acquisition and analysis. The daily ETC charging data will be stored in the basic data, and the traffic management department can retrieve and query at any time and form an analysis report.

5.6.3 *Accurate access to road travel data*

The road travel information is the actual overall operation status of vehicles in a certain section of the road which includes travel time, driving speed, etc. These data are important to evaluate the smoothness of road traffic and analyze the causes of vehicle delay. Especially in the big data management of expressway, the road travel data are the most effective and meaningful information. However, the traditional fixed detection method can only obtain the traffic flow information of a certain section, but cannot provide the overall travel data of the road accurately.

Using ETC system, we can normalize the time data of vehicles entering and leaving between the entrances and exits, calculate the time difference, and obtain the travel time data. By querying the road journey, we can get the data of the travel speed and the average travel speed.

By using these data, we can understand the characteristics of the evolution of expressway conditions and formulate the big data transportation management policies for expressway with target.

Of course, in practical application, because of the number of vehicles on the expressway in the ETC mode, it only accounts for a part of the total traffic flow of the expressway. Therefore, the total traffic flow can be calculated indirectly through proportional observation.

5.6.4 *Attempt to support charging to regulate traffic congestion*

In 2017, the second generation of the ETC system would begin to enter into practical application. Different from the first generation of the ETC system, the second generation system has faster interaction speed of road and vehicle information and more comprehensive data acquisition. Each data acquisition of the first generation is only 100–200 bytes at most, but after upgrading it to the second generation, the data volume will surge, and the communication speed will reach several K/s.

With this system, in addition to the regular non-stop charging function, data collection can also be used to support the regulation of urban congestion by charging.

For example, in the rush hour, when the data show that the traffic flow is close to saturation, the urban traffic management can issue instructions to increase the toll. When the traffic is underestimated, vehicles can pass for free.

Using the ETC system to acquire data and build big data transportation management system needs further exploration and attempt. From the adjustment and optimization of the internal structure of the traffic management department to the optimization of the channels for data acquisition and use, it should be unified. Only when the unified use and supervision are realized, can the ETC system play a greater role in a larger scope.

Chapter 6

The Platform of Big Data Transportation

The construction and development of big data transportation system is inseparable from the application of multiple platforms to big data. From single management module in the past to the upcoming smart transportation platform, and then to full integration into the intelligent city comprehensive management system ... it is inseparable from the connection and sharing of the big data transportation as well as the understanding and application of the characteristics of the traffic data from different sources.

6.1 Expressway Management Platform

At present, in the expressway traffic management, the construction of big data application management platform is in full swing. By integrating the existing business system data resources of the relevant departments of the expressway, we can build a network-based database, construct a data exchange platform, and establish a big data application management platform with rich data, advanced technology, and convenient use. In this way, we can promote the unified processing, exchange, and sharing of urban and regional expressway data resources, provide full data exchange and sharing guarantee for the whole expressway network, and bring diversified data services and support for managers and the public.

On December 20, 2017, Ningdu–Dingnan Expressway, the first intelligent expressway in Jiangxi Province, was completed and put into trial operation. The traffic flow can be monitored in real time when there are traffic control stations along the expressway. The monitoring system is set up in the tunnels and high slopes of key sections, which can accurately monitor the safe operation. In addition, BeiDou satellite technology is also used in this expressway to integrate the functions of mobile phone alarm location and road condition prediction so as to realize the rapid handling of traffic accidents.

As the first intelligent tourism highway in Jiangxi Province, the pavement reconstruction project of the provincial highway and Jiaowan highway has been completed recently. The highway is equipped with the functions of vehicle curve monitoring and warning, road icing warning, and automatic deicing.[1]

With the rapid development of expressway construction and the continuous expansion of the scale of road network, it is necessary to further break the obstacles of expressway management. The management organization should stand in the perspective of the whole road network, pay attention to the utilization of data resources, build a data management platform with rich data, advanced technology and convenient use and promote the unified processing, exchange and sharing of expressway data resources.

6.1.1 *Definition of the management requirements of expressway big data*

Only when the demand is clear can we have a definite target. To establish the expressway big data management platform, its demand points mainly focus on the following aspects:

(1) **Purpose of information sharing.** The basic and common data in expressway traffic information are sorted out to form a unified sharing standard so as to support different business applications.
(2) **Purpose of information summary.** The business data generated in the expressway traffic management is summarized to realize the transparency and scientificity of the business management of the superior and subordinate in the management organization.

[1]Reported by *Xinhua News Agency* on December 20, 2017, Reporter Gao Yiwei.

(3) **Purpose of information mining.** Big data management platform is used to find the management between different information data so as to achieve the deep use of data and improve the function and efficiency of management system.

6.1.2 *Overall framework design of the platform*

The expressway management platform consists of four parts:

(1) **Infrastructure level including two parts.** One is to support the necessary infrastructure and the other is to integrate the infrastructure. It includes the road section sub-center equipment and business interface equipment that can be included in the management platform and the optical fiber or network that connects the equipment.
(2) **Data resource level.** At this level, cloud computing technology is used to integrate the distributed physical resources and form the unified data resources. Through distributed database, the integration of hardware resources and data resources is realized.

    At this level, data resources are divided into three categories: basic data, business data, and video data. The first two are stored in the database and the video data are stored in the form of files.
(3) **Support platform level.** This level covers the software modules that support the underlying resources and upper applications, including application management platform software, distribution management software, and business process software. In addition, it also includes relevant system environment software.
(4) This level is oriented to end users and it is mainly applied to specific management business on the basis of support platform level.

6.1.3 *System data flow design*

In the whole expressway management platform, the system data flow is divided into four levels: data source, data processing, data application, and data presentation. After these levels, on the one hand, it can realize the integration of digital information resources and form a series of basic and business databases. On the other hand, these databases are analyzed and mined to form a comprehensive application, which is presented to users in different ways.

6.1.4 *Data interface system design*

Expressway management platform, as a centralized display platform of different data information, its core role needs to be reflected through the interface system. The interface system shall ensure the smooth connection of this platform with other business systems including the road administration management system, charging system, call center system, ETC system, equipment detection system, traffic detection system of traffic dispatching station, video monitoring system, etc.

The specific interface function modules are as follows:

(1) The external interface of networking data is mainly convenient for the third-party platform to retrieve the basic data for the expressway big data management platform.
(2) The interface module of road administration system can obtain the accident information, construction information, and hidden danger information of the regional expressway through the comprehensive road administration management platform of the regional expressway.
(3) The interface module of toll system can obtain the data information of total traffic volume, ETC traffic volume, passenger and freight classification, total traffic volume over the years, and so on.
(4) The interface module of the equipment monitoring system mainly obtains the working status of the equipment of the toll collection system and the equipment of the traffic dispatching station from the equipment monitoring system.
(5) The interface module of video monitoring system mainly obtains the monitoring video of expressway in the whole region.
(6) The interface module of meteorological system is used to receive regional meteorological data and information.

Of course, there are still many problems in expressway data management platform. For example, the system construction is self-contained, forming an information island, information resources not being shared, no access to traffic information for travelers, and so on. Whether in data acquisition or in data analysis and application, its ability of data processing and application needs to be improved.

It is reasonable to believe that in the near future, with the joint efforts of many departments and enterprises, the situation of cross-departmental cooperation and cross-system integration will be more significant. With the improvement of operation quality of expressway management platform, big data transportation management will create a good travel experience for the public.

6.2 Accident Detection Platform

Traffic accidents occur every day, which not only threatens the safety of travelers' lives and properties but also causes traffic jams and deterioration of road traffic conditions, leading to further increase of traffic accidents.

Therefore, it is necessary to strengthen the construction of urban traffic accident detection platform, which can not only deal with the traffic accidents on time but also make better use of big data technology, reduce losses, avoid congestion, and ensure the smooth flow of urban traffic network.

Hangzhou new city tunnel has established a perfect accident detection platform system. In the tunnel, a surveillance camera is set every 100 meters which can realize the whole process of seamless monitoring. These video monitoring systems are based on the extended application of the accident detection platform system. Therefore, no equipment is added on the site. Only a video signal is led out from the distributor of the CCTV monitoring system in the traffic management control room and connected to the accident detector.

After the processing of the accident detector, the platform can provide the corresponding traffic detection data and accident alarm signal to the central computer system in time. Therefore, the platform can automatically detect various traffic accidents including vehicle stop, traffic congestion, vehicle retrograde, speed exceeding the limit, object falling, traffic accidents, and other events. In addition, the platform can detect the fire and delay in a short period of time so as to provide accurate information for the monitoring personnel of the traffic management center to adopt effective measures to deal with the emergency and avoid greater losses.

Traffic accident detection platform refers to the use of a variety of equipment to operate together to collect incident data on the highway. Once the data information is found, that is to say, the accident occurs, the platform will automatically alarm and determine the point of the accident.

6.2.1 *Design of accident detection platform*

A complete automatic traffic accident detection platform should include hardware and software. The hardware includes traffic flow detector, communication facilities, and central computer for data processing; and the software includes data collection, processing, and data algorithm for accident detection.

6.2.2 *Operation mode of accident detection platform*

At present, the more effective operation mode of accident detection platform is as follows:

(1) **Finding sudden change data information.** After the traffic accident, the traffic data obtained by the platform will have a sudden change, mainly in traffic flow, occupancy, and speed.

For example, due to the occurrence of traffic accidents, it may cause a sharp decline in road capacity and form a traffic bottleneck. Through the real-time detection of the sharp reduction of traffic flow on the road section, the location of traffic accidents can be calculated.

After the traffic accident, it causes the vehicle to drive slowly, which results in the occupancy rate of the detector suddenly increasing. Besides, the occurrence of traffic accidents results in traffic jams, so that the speed detected by the adjacent detector is suddenly reduced. By acquiring and analyzing the aforementioned data, we can explain the abnormal traffic conditions on the road so as to measure and calculate the traffic accidents.

(2) **Finding data deviation from normal value.** The data information obtained by the detection platform deviates greatly from the historical information of the place and the time period in the past, which also indicates that there is a traffic accident on the road in the time period and the place.

For example, the traffic flow data detected by the detector is much smaller than the historical average value, indicating that traffic accidents may occur upstream or downstream of the detector. The occupancy rate detected by the detector is much larger than the previous historical data, indicating that there is a traffic accident downstream of the detector. The vehicle speed detected by the detector is much smaller than the historical data, indicating that traffic accidents may occur downstream of the detector.

The detection of sudden change data and data deviation from normal value is an important operation mode of traffic accident detection platform. In the actual operation of the platform, only one or two data changes cannot determine the traffic accident, and different kinds of data changes need to be combined to judge.

6.2.3 *Location of traffic accident*

The location of the traffic accident cannot be determined only by the data provided by the road section detector. After the data information is synthesized, the detection system can determine whether the traffic accident happened in the upstream or downstream of a section, but the distance from the section cannot be determined.

Therefore, the accident monitoring platform needs to integrate the data obtained by the adjacent section detector on the highway so as to determine the specific location of the traffic accident.

In fact, after the traffic accident, the data collected by the upstream detector and the downstream detector are different. Using this feature, the detection platform can determine the location of the traffic accident section as follows:

(1) **Data from upstream detector.** When the traffic accident occurs, the capacity of the road section at the location where the traffic accident occurs suddenly drops, forming a traffic bottleneck, which will cause traffic congestion at the upstream of the location where the accident occurs. Therefore, the traffic data detected by the detector will have the following characteristics: reduced speed, increased occupancy rate, and reduced traffic flow. By using these three data features, the traffic accident monitoring platform can judge that the location of the traffic accident is at the downstream of the location section of the monitor.
(2) **Downstream detector.** When a traffic accident occurs, the change of traffic digital information detected by the downstream detector is different from that detected by the upstream detector. Generally speaking, the changes of traffic flow digital information detected by the downstream detector have the following three characteristics: increased vehicle speed, reduced occupancy rate, and reduced traffic flow.

The increase in vehicle speed is due to the decrease in the traffic capacity of the upstream section of the detector after the traffic accident occurs in the upstream section of the detector, resulting in the decrease in the downstream traffic density. Therefore, the vehicle speed will naturally produce a significant increase. When the vehicle speed is increased, the number of share obtained by the downstream detector section or the number of share on the lane where the accident occurred, will be significantly reduced. Similarly, the data of traffic flow obtained by the downstream detector section will be significantly reduced.

In a word, by making full use of the function of data and collecting and detecting the traffic data information, we can diagnose the traffic condition on the road and find the time and place of traffic accidents in time.

6.3 Passenger Flow Detection Platform

Passenger flow data of traffic operation is an important data base of traffic planning, design, and operation management. Using the passenger flow detection platform to carry out scientific statistical analysis on the passenger flow data of urban traffic operation can provide a basis for the evaluation of traffic operation, assist and support the management department and the operation unit in formulating corresponding plans, and also provide data support for the planning and design of public transport.

In April 2017, Chongqing Traffic Administrative Law Enforcement Team released the news that Chongqing's first comprehensive traffic law enforcement closed management system was officially launched in Chongqing north railway station. The system can not only monitor the number of vehicles passing in real time but also reflect the demand of transportation capacity of North Station.

Four new eagle eye panoramic cameras added to the system can achieve 90% coverage of the ground monitoring of the North Square. More than 50 surveillance cameras are respectively installed on the first floor of the North Station, the exit of passengers, and other places so as to realize the coverage of the underground monitoring of the North Station. Through the surveillance video, the real-time situation of the scene can be understood.

The system can reflect the number of vehicles passing in real time. Eagle eye monitoring will clearly find the main gathering place of the crowd, especially in the peak period of the traffic flow in Spring Festival and National Day holidays. It can reflect the capacity demand of the North Station in real time so as to allocate the capacity in time to ensure the evacuation of passengers.[2]

The operation result of the passenger flow monitoring platform is an important basis for traffic operation and passenger flow data analysis. According to the location and time data of passengers and vehicles entering in and exiting to a certain area, the time and space rules of passenger flow can be analyzed to reflect the actual passenger flow demand.

Moreover, the traffic operation manager also needs to rely on the passenger flow monitoring platform to understand, analyze, and control the dynamic situation of traffic passenger flow under abnormal conditions, for example, the change rule of passenger flow in special weather conditions, holidays, and large-scale activities such as concerts and sports events.

Let us take the metro passenger flow detection platform as an example to demonstrate its application of big data.

6.3.1 *Design of passenger flow detection platform*

The function of passenger flow detection platform is to monitor the number of current passenger flow, obtain accurate data of passenger flow, and analyze it so as to realize dynamic and network management control of passenger flow, and then achieve comprehensive control of the passenger flow. Only when the passenger flow detection platform can fully and scientifically collect and use big data can it help the traffic management department to humanely disperse the passenger flow so as to effectively cope with the peak of large passenger flow and improve the operation efficiency.

A mature passenger flow detection platform consists of four parts:

(1) **Passenger flow monitoring module.** In this module, the number and flow direction of passengers in each region are obtained in real time,

[2]Hualong.com, April 18, 2017, Reporter Liu Yan.

and the dynamic monitoring of passenger flow is integrated with various monitoring technologies so as to maximize the accuracy of passenger flow data collection.

(2) **Control decision model.** According to the data obtained by the passenger flow detection module, different control decision schemes are generated. The establishment of this model depends on the passenger flow control index which maximizes the operation efficiency and passenger safety.
(3) **Simulation module.** According to the real-time data of passenger flow, it calculates the time needed for passenger flow evacuation so as to check the effectiveness of the decision-making scheme. After many times of simulation, the best decision-making scheme is obtained and stored in the detection platform.
(4) **Assistant decision display and information release software module.** It shows the results and controls decision-making scheme formed by platform detection to the management and control personnel and guides the passenger flow in different ways.

6.3.2 *Passenger flow monitoring at key nodes*

In the metro station and other large passenger stations, the flow process and route of passengers are referred to as "streamline". Taking the entrance as an example, from the entrance, security check device to the direct entrance (or purchase of temporary one-way ticket), then to the stairs, escalators, and metro vehicles, all are typical progress streamlines. The exit passenger flow is the opposite.

By monitoring each node, obtaining and analyzing the data, the platform can calculate the dynamic distribution of passenger flow in the stations.

Specifically, it can be divided into three areas: non-paid area, paid area, and platform area. At the key node of regional connection, the platform uses the monitoring means that fit the passenger flow characteristics of the location to collect data.

At the security check gate, two-way infrared technology can be used to detect passenger flow with infrared technology equipment. At present, the vast majority of security check gate equipment has technical functions. The passenger flow detection platform can develop an interface to connect the security check gate data record library with the central computer platform, so as to transmit the passenger flow statistics to the decision-making platform for processing.

At the gate, the gate data are used to count. In the original gate database, the data call port can be developed, and then the number of passengers recorded by the gate can be transferred to the detection platform for processing and analysis.

Further, video recognition and counting can be used at the stairway entrance between the platform and the station hall, and the electromagnetic induction technology can be used in the platform area, etc.

Different nodes and multi-technologies are used to detect the big data of passenger flow. The characteristics and laws of the streamline data of passenger flow are fully considered. The advantages of various technical means are made into full use. The data from different sources are centralized to the same platform. Such a decision-making platform for passenger flow detection and control can effectively reduce the workload of staff, improve the service level of traffic management and operation, and reduce the probability of accidents. In the future, the passenger flow detection system can be effectively integrated with other platforms and widely used, which has a good prospect.

6.4 Traffic Law Enforcement Platform

In recent years, with the rapid development of information technology, different law enforcement departments have established their own law enforcement platforms. However, due to the industry division and overlapping department share and functions of urban traffic management system in China, these law enforcement platforms still have relative limitations. Combined with the construction of big data transportation management pattern, the unification of traffic law enforcement platform is imminent.

In fact, in the process of transportation law enforcement, it is also necessary to interconnect with the information of public security, traffic police, and other departments. Among them, transportation, road, and administration law enforcement need to identify the authenticity of the information provided by the management object, which needs to be checked with the law enforcement data of the public security, traffic police, and other departments. Similarly, the public security and traffic police department also have a large amount of traffic video information. If they can be effectively connected to the unified law enforcement platform, these video information can also be used as a strong evidence for the work of transportation, road, and administration law enforcement.

In April 2016, Shanghai Jing'an Branch took the lead in the pilot application of traffic violation monitoring intelligent identification equipment, forming a new mode of traffic violation investigation and punishment combining "system intelligent collection and artificial screening". Subsequently, through the joint efforts of 16 districts and counties, the technology was promoted in Shanghai.

The system is mainly composed of four parts: front-end acquisition equipment, back-end identification, data transmission and exchange, and data interface, which integrates the image monitoring information collected by the front-end of the road, such as electronic police, public security checkpoint, and HD probe. This platform uses big data analysis technology to intelligently identify and analyze the illegal video data the traffic collected by HD monitoring equipment, and forms the illegal data of vehicle according to the technical standards. Finally, after being reviewed by the public security and traffic police department, the illegal vehicles are processed to realize the real-time detection and real-time alarm processing of traffic illegal.[3]

On the traffic law enforcement platform, the effective use of big data technology can discover the violations and locations of law enforcement counterparts in time so as to enhance the pertinence of law enforcement work and improve the efficiency of law enforcement. At the same time, the law enforcement behavior of law enforcement personnel can also be counted in time to reduce arbitrary law enforcement behavior, comprehensively analyze various law enforcement indicators, and improve the scientific decision-making level of law enforcement work.

6.4.1 *Data cloud center for traffic law enforcement platform*

To build a traffic law enforcement platform, a data cloud center must be established. The cloud center technology is used to integrate basic data information of road administration, transportation administration, taxi

[3]Reported by dongfang.com on July 15, 2016.

and other transportation industries including personnel information, vehicle information, household registration information, etc. In addition, it also includes the video information of key areas such as law enforcement business system information, highway intersections, surrounding of passenger transport stations, distribution centers of goods sources, inspection stations, etc. At the same time, it integrates the relevant video information and vehicle information of the public security and traffic police to build the big data cloud center for transportation law enforcement.

In the cloud center, we can make full use of the features of big data technology, such as mass storage and rapid calculation to realize the mutual exchange, full sharing, and mining analysis of traffic and transportation law enforcement information resources so as to support different application systems.

6.4.2 *Transportation integrated law enforcement coordination system*

On this platform, the auditing and law enforcement business systems of road administration, transportation administration, taxi, maintenance, and driving training are integrated and the law enforcement information of transportation management, road, taxi, maintenance, and driving training are also integrated. Through the sharing and connection of basic data and information among various law enforcement departments, the coordination of law enforcement and case handling of various departments is realized.

Moreover, the platform can also use big data technology to conduct in-depth mobile phone, mining, and analysis of monitoring video and the vehicle basic information in key areas related to traffic and transportation law enforcement cases. For example, by analyzing the regional distribution, growth, frequent location, source of vehicles involved, and other illegal behaviors of the case, we can get the distribution law behind the data. By monitoring the regional video, we can get more frequent and regular vehicle data, which provides support for law enforcement personnel to carry out accurate law enforcement. By analyzing the data of vehicles that violate the law many times in a certain period, we should establish a law enforcement system that focuses on monitoring.

6.4.3 *Traffic law enforcement information service system*

This system can be publicized through the multiple means of website, big screen, Weibo, WeChat official account, and mobile phone APP so as to publicize the information related to traffic law enforcement cases. By using this system, different departments can query the case information and law enforcement personnel, and can also realize the public's online complaints and reports of irregular law enforcement behaviors so as to improve the information transparency of traffic law enforcement (see Figure 6.1).

In addition, through this system, we can actively push and exchange the illegal data information and share it with taxi, public transport, and other enterprise managers, so as to realize the coordination of law enforcement and enterprise management.

No matter if it is the traffic police, road administration, or the transportation administration and taxi, in the process of traffic law enforcement, they need to interact with the data of other departments. Therefore, it is imperative to use the big data technology to establish a

Figure 6.1 Intelligent traffic surveillance system.

Source: Security Website, Shanghai's first Intelligent Traffic Surveillance System.

unified and comprehensive traffic law enforcement platform and integrate business data.

6.5 Logistics Information Platform

With the continuous improvement of mobile Internet technology and the advent of big data era, the social life and consumption patterns have changed dramatically. Using modern big data transportation management technology can reduce the cost of enterprise operation and improve the efficiency of enterprise production and operation. Therefore, the logistics information platform must be effectively combined with big data information technology to make the innovation ability of logistics management rise to a higher stage.

The "Qinglong" big data logistics operating platform self-developed by Jingdong Mall can collect customer information at the front end of the system, while the back end corresponds to the company's logistics department and the third-party logistics company. The "Qinglong" platform realizes the automatic goods sorting system by using big data technologies such as deep neural network and map area division so as to ensure that customers receive goods in the fastest time.

Alibaba Group's Cai Niao Post Station has also established a similar logistics information platform. On the platform, through the use of big data technology, the in-depth investigation and planning are conducted to determine the specific location and quantity of the post station.

Using similar big data logistics information platform can ensure the economic benefits of logistics service providers and users, and also make it more convenient for customers to send and pick up goods.

6.5.1 *Improving the operation function of logistics information platform with big data*

By using big data technology, the management and control system of logistics information platform has changed greatly. Among them, the most outstanding innovation is the early warning function.

In the process of logistics operation, the receiving, storage, transportation, distribution, maintenance, and management of materials are indispensable from the collection and processing of logistics information. The use of big data technology on the platform can effectively improve the accuracy of market demand forecast, and can also give early warning on the status of product inventory and transportation.

For example, if a product is placed in the warehouse for more than a period of time, big data technology can play a role and send an early warning signal on the logistics information platform. In this way, employees at relevant positions can handle it in time.

In addition, the introduction of big data technology can also help enterprises and traffic management departments share data and pay real-time attention to the transport status of transport vehicles. For example, once a vehicle encounters a traffic accident or component damage, it can be found in time so as to eliminate potential safety hazards in time.

Therefore, using big data technology to improve the operation function of logistics information platform can effectively improve the efficiency of logistics operation and improve the economic benefits brought by logistics.

6.5.2 *Strengthening the ability to use big data information*

In the process of data integration with big data, the logistics information platform needs to discover and collect data related to logistics business. It includes the customer's requirements for product order quantity, product inventory, best transportation route, packaging, distribution and other related data information.

Among them, the collection of these big data information is the premise to realize the efficient operation of logistics information platform and the data information must be scientific, timely, and complete. At the same time, the data in the era of big data is not only massive, but also constantly changing. Therefore, the logistics information platform should increase the information storage room not only by increasing the number of traditional information storage equipment but also by relying on network storage system, virtual storage technology and virtual data storage system.

In the process of calculation and analysis, the logistics information platform should be realized by using the integrated machine database, data mart, and other commercial big data analysis tools. In this way, data and information can be processed and analyzed scientifically to ensure that they can be used as available data and information resources.

6.5.3 *Improving the security level of logistics information platform*

We must pay attention to the fact that the logistics industry is one of the industries that hackers mainly focus on, and the leakage of data information will obviously bring loss risks to logistics enterprises and customers. Therefore, in the big data transportation environment, the logistics information platform must pay attention to the following aspects of data information protection:

(1) The centralized storage of big data information is bound to increase the risk of theft;
(2) Due to the massive storage characteristics of big data, there must be a large number of invalid data inside, which may cause the risk of making malicious error information;
(3) The data in big data logistics information platform has a wide range of information sources, so it is necessary to expand the scope of security protection.

In view of the aforementioned problems, the logistics information platform can improve the technology for security protection system, realize the effective screening, decentralized storage and classification management of data information, and better protect the security of the logistics information platform.

Under the big data transportation environment, the establishment and construction of logistics information platform need further research and analysis. We should fully not only find the combination point of the big data technology and traditional logistics management but also solve the problems such as imperfect similar operation skills, inextensive use of big data, or security protection system.

It is reasonable to believe that when the logistics information platform makes full use of big data technology, the level of logistics industry management will be able to achieve a new leap.

6.6 Emergency Command Platform

The advent of the era of big data transportation has brought multiple meanings to the government emergency management. On the one hand, in the traffic emergency command, the management department must discard the false and retain the true data from different sources on the platform and verify the comprehensiveness and credibility of the data from multiple perspectives, which lead to the complexity of the emergency command platform to collect and process big data. On the other hand, the application of big data also brings good prospects for the management of emergency command platform.

On January 6, 2016, Weifang Transportation Emergency Command Center was officially put into use. The emergency command system not only realizes the video integration of the key traffic places in the city but also integrates the information dispatching management system of taxies and the dynamic management system of road transport vehicles and realizes the interconnection with the public security organs. The platform can be used to supervise whether the security check in the station is normal and whether the public needs to call the vehicle trajectory when picking the lost items in the taxi.[4]

On the emergency command platform, competent departments can continuously receive transportation-related data through satellites, sensors, video surveillance, mobile communications, radio frequency identification equipment, social media, etc. Through the intake, analysis, distribution, and function of these data, they can be transformed into meaningful and valuable crisis management information.

Therefore, in a crisis situation, the transportation authorities can manage, model, share, and transform the big data appropriately so as to make emergency decisions in a way that is unimaginable in the traditional situation.

---

[4]Reported by *Weifang Evening News* on January 7, 2016.

6.6.1 *Sharing of big data by emergency platform*

In the era of big data, the integration and sharing of big data information has become the most basic requirement for the construction of emergency platform system. The management department should take the initiative to change, break the original data separation and blockade, and open the information. Based on this, the construction of emergency platform system in big data technology can improve the efficiency of different government agencies and departments.

As for the construction of emergency platform system, the technical system can be built based on big data technology including wired communication, wireless communication, mobile command communication, mobile Internet technology, and big data application software, which have different functions such as wired communication dispatching, wireless communication command, mobile emergency command, and remote consultation. Through the government intranet, extranet, and Internet, we can build a unified information platform for public security emergency management of traffic emergencies. In this way, the traffic management department, together with public security, water affairs, metro, electric power, heating, gas supply as well as other enterprises and institutions, can get comprehensive urban operation monitoring data, which provides important support for making scientific emergency command and decision.

6.6.2 *Cloud emergency system*

In the era of big data, cloud computing technology is an indispensable IT foundation. The emergency platform shall establish a cloud emergency system based on cloud computing technology to provide intelligent emergency services for emergency departments.

Specific to the traffic emergency information platform, the cloud emergency system should be able to collect scene perception data, historical data, and decision support model for real-time traffic emergencies. These data are seamlessly integrated through standardized interfaces to achieve rapid and comprehensive mining and extraction and realize the comprehensive integration of information and data on the cloud emergency platform. According to different emergency needs and the service nature of different departments, an emergency command collaborative network is formed.

6.6.3 *Application of mobile monitoring and emergency response*

In case of natural disasters, traffic accidents, or the local monitoring equipment being difficult to work, the traffic monitoring center can quickly understand the site situation, make objective analysis, and rescue treatment by relying on the emergency command platform.

Depending on big data technology, the specific scheme of mobile monitoring and emergency response is as follows:

(1) The mobile platform quickly mobilizes to the site to carry out work. It includes opening satellite channel, connecting video phone system with remote emergency command center, and returning field information to detection center in real time.
(2) Internet channels are used such as websites, Weibo, WeChat, SMS, and other carriers to timely release emergency road conditions, weather warning reminders, and suggestions for detours.
(3) According to the report of the mobile platform personnel, the personnel of emergency command platform take the available monitoring equipment around the accident, pay close attention to the surrounding traffic flow change, and report the relevant situation to the higher supervision department immediately after finding or receiving the relevant abnormal situation.
(4) After receiving all kinds of on-site data, the emergency command platform personnel shall collect and sort them out in time to study the future trend of time. Through data sharing, they shall coordinate the road administration, traffic police, toll collection department with other departments, carry out rescue or traffic guidance, and do a good job in flow balance control so as to stop all potential disaster events in the first time.

The emergency command platform plays an irreplaceable role in improving the means of big data transportation supervision, improving the efficiency of emergency response, strengthening the traffic management and control and promoting the linkage of various departments. Strengthening the operation of big data can improve the work level of the emergency command platform so as to improve the operation and management level of the traffic management department.

Chapter 7

Big Data Transportation and Logistics

As a big trade country, the logistics in China is used more and more frequently, which actively promotes the economic growth and development in China. As a strategic industry, the logistics is widely concerned by all walks of life, and the arrival of big data transportation can provide rich and effective practical information for logistics information. In order to make the logistics industry get better development, it is very important to study the establishment and construction of the relationship between logistics and big data transportation.

7.1 Big Data Transportation and Transportation Organization

Logistics transportation organization includes a series of transportation activities, such as investigation and prediction of transportation, preparation for transportation plan, allocation of transportation vehicles and ships, organization of passengers and freight stations, transportation vehicle scheduling and route optimization, tracking, positioning and monitoring of transportation process, and coordination of multimodal transportation.

The organization of information transportation is inseparable from the application of big data transportation technology, which permeates all aspects of transportation production including the application of electronic technology, network technology, and communication technology, and has different forms of software and hardware.

In December 2017, CCTV1 focused on the drone delivery of JD, which prompted many people to pay more attention to the new transportation organization system.

At present, JD has built the world's first UAV operation and dispatching center. Through big data technology, the coordinate position and flight parameters of each UAV can be seen clearly on the monitoring large screen. By monitoring the route and order data, the staff can guarantee that the cargo can fly to the destination according to the voyage set by the system.

Through the UAV transportation service supported by big data, it used to take two or three hours to deliver ordinary daily necessities, but now it only takes roughly 20 minutes.

By using big data transportation technology, we can fully improve the logistics efficiency, help enterprises to improve the problems, and expand the promotion space by optimizing the transportation organization.

Specifically, the impact of big data transportation technology on transportation organization is reflected in the following points:

7.1.1 *Transportation environment forecast*

In the past, logistics enterprises were always used to predict the transportation environment through past experience, questionnaire, or prior contact, for example, looking for customer sources, designing the best route and understanding the traffic changes. But in fact, when the findings are summed up, the transportation environment is likely to change.

Compared with the past, big data can help logistics enterprises to fully and accurately outline the transportation environment and reflect the environmental changes through real and effective data so as to make predictions on the various stages of product transportation. In this way, the transportation organizer can reasonably set the transportation plan.

7.1.2 *Location of transportation nodes*

In transportation organization, the location problem of transportation nodes requires logistics enterprises to minimize the distribution cost on the basis of fully considering their own business characteristics, commodity characteristics, and traffic conditions. This problem can be solved by using the classification tree method in big data technology.

7.1.3 *Optimization of transportation routes*

In transportation organization, logistics enterprises can use big data to analyze the specifications and characteristics of commodities and the different needs of customers in terms of time and money, and then use the fastest response to accurately reflect these factors that affect transportation plan, for example, which transportation scheme to choose, which transportation cost mode to adopt, which transportation route to choose, etc. In this way, the most reasonable transportation route can be made.

At the same time, the logistics enterprises can analyze the traffic situation of the transportation route quickly through the real-time data received in the distribution process so as to give early warning to the situation that may affect the transportation route, accurately analyze the information of the transportation process, and make the transportation distribution management more intelligent.

7.1.4 *Bin-location optimization*

It is of great significance to arrange the bin-location of commodities reasonably for improving the efficiency of transportation route management. For those transportation routes with large quantity of commodities and high frequency of delivery, optimization of bin-location brings direct work efficiency and economic profit.

In fact, which commodities can be put together at a constant speed to improve the sorting rate and which commodities have short storage time and high transportation requirements can be analyzed through the big data association mode method so as to get the mutual data relationship of commodities in the transportation process to arrange the transportation location more reasonably.

The most essential difference between modern logistics and traditional logistics lies in transportation organization. Modern logistics cannot do without the support of big data and other information technology, which is reflected in the transportation organization with systematization, informatization, networking, and intelligence.

Currently, China's logistics enterprises have not fully integrated the transportation organization with the functions of packaging, storage, circulation and processing, delivery and information processing because of less computer application, less awareness of big data application, lower quality of employees, and so on. Therefore, it is of great significance to pay attention to the role of big data in transportation organization to improve logistics management decision-making.

7.2 Big Data Transportation and Rural Logistics

Compared with urban logistics, rural logistics has more levels of transportation. Rural logistics is usually carried out in three steps, as warehouse tallying, regional transportation, and the last-kilometer distribution. In fact, in the vast majority of rural areas, except for the last kilometer of distribution, other logistics steps are similar to urban distribution. The last kilometer of distribution has become the most critical step to determine the efficiency of logistics.

At the end of May 2016, at the 3rd China County E-commerce Summit, Xiong Jian, head of rural logistics of Cai Niao network, said that Cai Niao network is using the data system to organize the transportation capacity of express delivery and logistics companies around the country and build a more dense and deeper distribution network in the rural areas. Through the rural terminal logistics network, more than 30% of rural parcels can be delivered from the county to the rural area on the same day, and nearly 70% of parcels can also be delivered to the village the next day. Cai Niao network has established a rural terminal network covering more than 320 counties and nearly 16,000 rural service stations in the country.

Due to the fact that the address of the consignee is not clear, the path planning is not in line with the reality and there are too many accidental factors in the rural logistics, common phenomena such as wrong delivery and multiple delivery in the logistics, resulting in the increase of the delivery cost and the decrease of the satisfaction of the consignee, etc. In view of the problem of "last kilometer" delivery of rural e-commerce logistics, many logistics companies have tried to use a variety of special forms of delivery, but have not effectively improved the efficiency.

In view of this, we must introduce big data transportation technology, correctly analyze the regional distribution state of rural logistics and the living conditions of the consignee, and put forward targeted methods to effectively solve the rural logistics problem.

7.2.1 *Innovation of transportation service*

The collection, sorting, analysis, and application of big data can effectively integrate the social transportation capacity of rural logistics so as to improve efficiency and reduce costs. The key to the integration of rural logistics resources lies in the establishment of a special logistics information platform, which can be carried out through the following two modes.

Large enterprises shall take the lead to establish a special rural logistics information platform and be responsible for its operation. When other logistics enterprises or individuals apply it, the data of each business and shipment generated by them will be recorded. Through the analysis of a large number of data, consumption habits, credit records, address accuracy, path credibility, and so on in rural areas will be recorded and displayed so as to provide an effective basis for later customer selection.

A special transportation capacity ecosystem shall be built. Through the installation of GPS equipment for rural freight vehicles, the logistics companies may have thousands of vehicle data. Although these vehicles are different in functions and models, they can form the basis of rural logistics capacity. In addition, large-scale logistics can also establish a private information platform, master and integrate their own information so as to provide personalized and professional logistics information services. When the aforementioned platforms are connected with each other, a logistics capacity system covering the rural market of a certain region can be formed. For example, in the face of the rural market, if the private transportation capacity of an enterprise is insufficient, it can be obtained from the social transportation capacity. In the same way, social transportation capacity can actively serve enterprises in need according to its own characteristics. Especially in the role of big data technology, vehicle and goods sources can ensure timely and reliable information, which is conducive to the last kilometer of freight transaction in the rural market.

7.2.2 *Solving the problem of insufficient transportation capacity of the "last kilometer"*

The use of big data can solve the problem of insufficient transportation capacity in the last kilometer of rural logistics delivery. Logistics enterprises can develop an open information system for the public and rural workers, students, grass-roots government personnel or long-distance freight drivers, taxi drivers, and so on, through their mobile applications, to undertake logistics delivery tasks. In this way, enterprises can obtain "temporary transportation capacity" and manage it effectively so as to reduce the load of their own transportation capacity.

This kind of crowdsourcing model based on big data is especially suitable for promotion in rural areas and mountainous areas with a large population. Of course, in the face of such a large number of management of mobile logistics transportation resources, we must have massive data processing capacity of the big data technology to complete.

7.2.3 *Speeding up the construction of rural logistics information network*

The application of big data technology in the field of rural logistics is inseparable from a strong rural logistics information network. Such a task cannot be completed by any single enterprise and it must be coordinated and gradually completed by relevant government departments.

On the one hand, it is necessary to establish and improve the agricultural product information center, establish a responsive agricultural product information network and ensure that big data can be discovered and transmitted in a timely manner. Especially in the case where the whole rural information network is still being established, the infrastructure of big data collection of agricultural products must be further improved.

On the other hand, it is necessary to train rural logistics professionals. Through the cooperation among enterprises, government departments, and research institutions, a variety of educational resources shall be used, and through different forms and levels of training, the quality awareness and industrial awareness of logistics service, and managerial staff shall be improved to use information technology to comprehensively improve the level of big data information service in rural areas so as to meet the needs of talents in rural logistics service.

7.3 Big Data Transportation and Urban Delivery

At present, China's urban logistics lacks overall system planning, and the routes for urban logistics delivery are in chaos. In addition, urban traffic restriction, traffic congestion, and diversified customer demand lead to relatively low efficiency of urban logistics at this stage. Fortunately, with the rapid development of Internet and information technology, big data transportation will be able to change the traditional urban logistics industry.

On the morning of December 16, the first public logistics information platform of Yiwu City, Dunda Delivery, was officially put into operation. Dunda Delivery is the urban public logistics information platform independently developed by Yiwu Dunda Supply Chain Management Co., Ltd. The platform supports online order placing, online order receiving, online settlement, online query, and other functions to realize the whole process of information management from delivery to signing in. It implements the unified control of distribution vehicles with clear operation track, and fully ensures the timeliness and safety of goods delivery. It also reasonably incorporates social vehicles, realizes the sharing delivery and improves the vehicle utilization rate and real load rate.[1]

The lack of overall planning of urban logistics will cause repeated investment and huge waste of logistics facilities of different enterprises and restrict the effective utilization and integration of resources. At the same time, each logistics enterprise, based on their own interests, is confident in planning the logistics center and delivery route and does not share the logistics data, so it is unable to form an optimal delivery system at the macro transportation level. In addition, the development of mobile Internet also makes the demands of urban logistics customers tend to be customized. Many customers want to be able to grasp logistics information and control all aspects at any time from many aspects and angles.

In the face of these situations, big data technology should be used from the following perspectives to solve the problem.

7.3.1 *Government takes the lead to establish the urban logistics resources sharing platform*

Under the leadership of the competent government departments, the city logistics resources sharing platform will be built. The government should allocate local logistics resources effectively and monitor them in the

---

[1]Reported by *Yiwu Business Daily* on December 17, 2017, Reporter Wu Fengyu.

process of urban delivery. In this way, it can not only realize the local economic growth, but also make full use of the big data obtained in the logistics delivery to realize the integrated planning and management of urban logistics and even more industries. Of course, for the government, it is quite a challenging goal to directly seek the cooperation of urban logistics enterprises and complete the extraction, collection and management of logistics big data.

7.3.2 *Industry alliance to build a sharing platform*

Under the leadership of the government, an urban logistics delivery industry alliance involving logistics companies and enterprises can be established. This is because the logistics industry enterprises can master the first-hand information of urban logistics and the industry alliance can build the information resource sharing platform, which can solve the source problem of logistics big data. At the same time, the industry alliance stores all the data and information in the same database to facilitate the instant sharing and information exchange between logistics locations and avoid the traffic jams, disordered logistics operation, low customer satisfaction, and other problems.

7.3.3 *Attaching importance to the construction of urban logistics and storage platform*

In order to build a sharing platform of urban delivery resources, it is necessary to use big data to support the urban logistics storage platform. Urban logistics storage platform is an important factor affecting the efficiency of urban delivery, whose foundation includes storage center, dispatching center, settlement center, etc. Through this platform, all delivery channels are comprehensively analyzed, including human resources, material resources, item flow direction, delivery speed, and delivery cost so as to calculate the best location and delivery plan and optimize the resources (see Figure 7.1).

In a word, the best way to solve the problem of urban delivery is to focus on the big data transportation technology led by government departments, participated by logistics enterprises, provided technology by professional companies and build a platform for sharing urban distribution information.

Figure 7.1 Logistics truck.

7.4 Big Data Transportation and Express Logistics

Currently, express logistics has entered every corner of people's life. It is no exaggeration to say that many times, the quality of an express delivery will affect our life experience. Facing the continuous upgrading of e-commerce industry, express logistics enterprises also need to actively transform and upgrade, using big data technology from homogeneous competition to diversified competition.

On May 31, 2017, the Internet conflicts of Shunfeng and Cai Niao network under Ali became public. Cai Niao issued a statement claiming to have received a notice from Shunfeng about the suspension of data interface of Fengchao. Subsequently, Shunfeng successively closed the data interface of Fengchao self-delivery cabinet for Cai Niao and the pass back logistics information for the whole package of Taobao platform. On June 1, Alibaba announced that it recommended that merchants shall suspend the use of Shunfeng shipping.

However, the statement issued by Shunfeng is totally different. In May, Cai Niao stopped their cooperation with Fengchao for the sake of data security and offline the Fengchao interface information from 0:00 on

June 1, and then Alibaba Group platform removed Shunfeng from the logistics options.

Why do the giants of e-commerce and express industry have such conflicts around data? In fact, the outbreak of commercial contradictions reflects that Chinese express logistics enterprises are likely to complete the integration and promotion of the industry inside and outside by virtue of the value of data resources in the era of big data transportation.

In fact, although the business volume of China's express service enterprises is rapidly increasing, there is an inevitable crisis under the seemingly good situation. Due to the transformation of the competition mode from scale war to price war, coupled with the height of labor cost and the gross profit rate of express delivery business begins to fall sharply, while facing the cross-border enterprises from e-commerce, such as JD and Suning. Therefore, express companies must take advantage of big data resource allocation as soon as possible to expand their competitiveness and improve market influence.

7.4.1 *Use of e-waybill in express logistics*

E-waybill is a kind of information waybill with high efficiency and environmental protection. Different from the traditional express delivery's triple sheet and quadruple sheet, the e-waybill is printed by computer with two-dimensional code logo and the adhesive on the back makes it easy to paste and tear.

With the e-waybill, the express delivery system can easily identify, process, and deliver a package among hundreds of millions of packages. Through the data flow supported by the e-waybill, the enterprise's own e-waybill system can automatically connect all the data information of the consignor, the express company, the consumer, and the main and branch lines. According to the large amount of big data connected, the enterprise can optimize the express link.

At present, most of the mainstream domestic express companies have completed the popularization of e-waybill, which shows that the big data technology has become the standard configuration of express logistics enterprises.

7.4.2 *Application of big data routing sorting*

Big data routing sorting can directly complete the accurate matching of packages and outlets through the big data analysis for massive addresses and the combination of spatial positioning technology of Internet map. At present, the application accuracy rate of this technology is more than 98%, and with the big data precipitation of each express logistics company, the accuracy rate will soon be close to 100%.

Before adopting this technology, a large number of parcels from all over the country must be concentrated in the distribution center of the enterprise, then classified according to the shipping address, and then distributed to the next outlet. In this process, there will be a large number of sorters on the assembly line. They have to look at the address information on the package, and then use their memory to determine which outlet the package will arrive at next. With the big data routing sorting technology, the sorting speed can be reduced to 1–2 seconds per order and the warehouse sorting efficiency is greatly improved.

Currently, the big data routing calculation can ensure that after the order is generated, it can be displayed to the delivery outlet soon. The popularization of this technology can help express companies to forecast the delivery volume of outlets.

Moreover, according to the four-level address library generated by big data routing calculation, the express companies can match the delivery address of consumers to the structured towns and streets in the first time. With these structured address information, the express company can accurately locate the receiving and delivery addresses so as to provide more accurate delivery and route planning for the courier.

7.4.3 *Big data technology for accidental circumstance management*

Most of the operational pressure of express companies comes from accidental circumstances. Among them, the worst accidental circumstances include the abnormal packages that cannot be delivered normally and the period when the delivery pressure skyrocketed.

Over time abnormal packages are the express packages that have not been delivered within more than 48 hours. Through the big data technology, these packages and order data can be effectively screened so as to help express logistics companies to timely understand how many overtime abnormal packages they have generated and which outlet is the most serious. Through the order data, we can know the reason in time so as to start to improve pertinently.

Package volume skyrocketed is usually concentrated in the peak periods of e-commerce consumption such as holidays, Double Eleven, and Double Twelve. Through the logistics early warning system, it can predict the package volume in advance using big data so as to guide e-commerce merchants to prepare for warehouse delivery and help express companies to allocate limited capacity resources.

In addition to the aforementioned applications, the big data technology of express can also help e-commerce companies control the final process of online shopping. Through the whole process monitoring of the circulation data of logistics orders, the e-commerce platform manager can obtain the data, identify the false orders such as "credit swiping" and purchase praise" so as to realize the effective management of the merchants.

At present, big data has been widely used in express business. In the future, with the growing maturity of big data transportation environment, the technology will fully penetrate into every link of express business more and more and become an important part of express logistics industry.

7.5 Big Data Transportation and Cold Chain Logistics

With the continuous development of logistics industry, along with the continuous progress of science and technology and the rapid growth of refrigeration technology, the cold chain logistics based on freezing technology as a basis and refrigeration technology as a means, is increasingly valued by the market. Due to the continuous enrichment of material life and the increasing attention of consumers to food safety, the cold chain logistics is widely used in the delivery of agricultural products, meat, and other products.

On the other hand, China's current cold chain logistics system had not developed fully. In the process of logistics distribution, the problems of high loss rate and low circulation rate have not been completely solved. How to apply big data technology to cold chain logistics and realize the integration of cold chain logistics so as to develop and expand the cold chain logistics industry has become an urgent task.

On September 10, 2016, China Cold Chain Logistics Intelligent Cloud Platform was officially launched in Zhenjiang, marking that the first domestic cold chain logistics intelligent cloud platform, based on big data and Internet of Things technology and integrated online display transaction, cold chain logistics order, cold chain logistics warehousing and other functions, is into the market operation.

The intelligent cloud platform can realize intelligent monitoring of cold chain logistics, online order matching, and online order trading. The platform will effectively integrate the upstream and downstream information of the industry and realize the whole process monitoring of cold chain transportation. Zhejiang data monitoring center can monitor the temperature, humidity, GPS location, and other information of cold chain transportation in 24 hours. Once there is an exception in the transportation process, it will receive an alarm at the first time and conduct human intervention at the first time, which greatly enhances the safety of cold chain transportation.[2]

At present, many small and medium-sized cold chain logistics enterprises operate independently, the business model is relatively scattered, the product and logistics information is not smooth, and the transmission of information is more hindered. Compared with the developed countries, the application level of big data technology in the field of cold chain logistics in China is relatively low, which is still at the primary stage.

Specifically, the lack of data collection and system is a serious problem in the application of big data technology in cold chain logistics. In addition, the data collection of agricultural products transported by cold chain logistics is relatively independent in the market and government departments, relying heavily on traditional manual operation and lacking automatic collection methods and means, so the accuracy, objectivity, and timeliness of data are not strong.

Therefore, it is necessary to promote the combination of big data transportation technology and cold chain logistics industry so that logistics enterprises, cold chain product suppliers, consumers and government regulators can view transportation information, product information, consumer information, and weather information through big data.

The specific methods are as follows:

(1) **Using big data technology to establish cold chain product traceability system.** Most of the products transported by cold chain are agricultural products in a broad sense, including fresh products, fruits

[2]Reported by guangming.com on September 12, 2016.

and vegetables, crops, etc. It is necessary for the cold chain industry to establish a big data platform for information traceability, to use Internet of Things technology to collect data sources, and establish an intelligent traceability system. In this way, a complete intelligent information tracing network can be formed to ensure the quality and safety of cold chain products in transportation.

For example, cold chain products should be equipped with location data, manufacturer data, production cycle data, delivery data, etc. In this way, the circulation of cold chain products can be managed with the big data in the whole process.

(2) **Using big data technology to comprehensively coordinate cold chain logistics and transportation.** At present, many areas have used monitoring technologies such as Internet of Things, GPS, and intelligent vehicle terminal to improve the efficiency of cold chain logistics comprehensively through the mobile phone and processing of big data.

For example, the establishment of vehicle identity and authentication system can reduce the operational risk of cold chain delivery center and the third-party logistics company. Another example is to realize the whole process visual monitoring of cold chain delivery vehicles, from the running speed of vehicles, product status to the driving route and driver status to realize the whole process visual monitoring so as to realize the reasonable allocation and management and supervision of vehicles.

More importantly, the cold chain logistics transportation system can monitor the vehicles in real time through GPS to ensure that it can check whether the vehicles have smoke, whether the temperature and humidity can meet the requirements of product storage and transportation. Through the intelligent onboard system, it can also transfer the technical parameters of refrigerated vehicles such as documents and drawings to the monitoring center of the enterprise in real time. Whether the product consumers or enterprises, they can view the relevant parameters through the Internet.

(3) **Guaranteeing timely and accurate operation strategy adjustment of the background.** The introduction of big data in time can ensure that cold chain logistics enterprises can obtain integrated real-time data at any time when they study strategies, and obtain accurate and fast query through the inquiry and processing of unstructured data.

At the same time, in the cold chain logistics industry, the most important feature is the timeliness guarantee. Due to the special nature of products delivered by cold chain logistics, such as the uneven product source, the typical seasonality of various products and the peak and valley years of production, the costs and risks of cold chain logistics enterprises have greatly increased. Using big data technology, we can adjust the operation strategy of cold chain logistics enterprises through the existing data distribution law and support them from the internal and external. In addition, enterprises can get the most needed data in time to define the deployment plan that needs to be accurate including obtaining maintenance cost, failure frequency, operation route, etc.

It has become an inevitable trend for the existing cold chain logistics enterprises to integrate cloud computing, big data, mobile Internet, and other technologies for cold chain logistics. It can be confirmed that the application of the big data technology in cold chain logistics has infinite development potential. Big data transportation service and cold chain logistics will go to a broad development prospect together.

7.6 Big Data Transportation and Multimodal Transportation

The multimodal transportation refers to the process of logistics and transportation completed by two or more means of transportation linked and transshipped with each other. With the development of national economy, the vigorously developing multimodal transportation is the key work of comprehensive transportation development during the 13th Five Year Plan period. Strategically, it is an important starting point for China to promote the supply side structural reform of transportation. From the perspective of enterprises, it is an effective way to reduce logistics costs.

In the process of innovation of the development strategy of multimodal transportation, at present, all regions have jointly presented an important leap toward big data transportation.

On May 19, 2017, the international logistics multimodal transportation data transaction service platform of Zhengzhou Airport was officially launched. The launch of the platform realized the information sharing of domestic road, railway, seaport, and other transportation modes with the entire transportation chain of international aviation and overseas land transportation and ensured the formation of information linkage in Zhengzhou Airport, free trade test zone, bonded park, and other areas.

In terms of the government, the relevant departments in Zhengzhou said that they would provide one-stop information services to achieve "one stop" of information services for logistics enterprises. In addition, they will rely on leading backbone enterprises to build an Internet data interaction platform involving multimodal transportation enterprises so as to continuously release the dividend of big data resources.

In the process of multimodal transportation, building a big data platform is to be equivalent to implanting "central nerve" for all participants of multimodal transportation. On the multimodal transportation integrated information platform that has been built and applied, the big data transportation technology has established branch systems including booking information platform, box management information system and highway logistics system, which can realize the whole information monitoring of vehicles, boxes, and goods.

7.6.1 *Application space of big data in railway supply chain*

The railway freight transportation plays an important role in the multimodal transportation. The support of big data technology to the transformation of railway supply chain also needs to expand more space. In fact, the Internet, big data, cloud computing, and other scientific and technological means can enable railway freight transportation to find new opportunities for improvement by virtue of its original advantages.

Currently, the railway freight transportation only plays the role of carrier in multimodal transportation, and its advantage in attracting customers lies in low freight cost and long distance from terminal customers. The ability of railway freight transportation to use other logistics transportation channels is not enough, and it cannot share data with the resources of road, water, and air transportation smoothly.

Therefore, the railway freight transportation must know how to use the advantages of big data, and construct its own unique data assets through the effective promotion of container multimodal transportation. Then, according to the operation mode of multimodal transportation, the data sharing and integration of upstream and downstream logistics enterprises are realized and the cross-border integration is completed. Finally, we can realize the effective realization of big data resources.

7.6.2 *Optimization and scheduling of multimodal transportation routes*

In the process of logistics delivery of multimodal transportation, the most important thing for enterprises is the product quality and transportation speed. In order to ensure the product quality and reduce the transportation cost, the enterprise must optimize and schedule the route continuously.

Through the induction system inside the multimodal vehicle, when loading and unloading products, the onboard system can upload the data information of the sensors on the products to the server. After receiving the data, the server can analyze the vehicle and product information in real time, and carry out dynamic calculation according to the actual transportation capacity, customer preferences, and delivery route data so as to get the latest and optimal route. With the precipitation of data and the progress of technology, the real-time update can be made according to the GPS positioning system in combination with the traffic conditions.

From a macro point of view, the remote control system can also get the remaining delivery time in the process of multimodal transportation through big data analysis so as to timely inform customers to pick up the goods. Because the multimodal transportation is a continuous process, the chain break is easy to cause the failure of the whole system. Therefore, we must make use of big data to realize the overall linkage from the manufacturer to the customer, reduce the damage rate to the minimum, minimize the waiting time, save the cost of the logistics company, and improve the delivery efficiency.

7.6.3 *Improvement of container multimodal transportation promoted by big data*

The standardized container is the basic unit and operation basis of multimodal transportation. However, in China's highway transportation and railway transportation, the utilization rate of containers is quite low, restricting the development of multimodal transportation.

In view of the current difficulties of the container multimodal transportation, the container unit segmentation will be able to adopt the transportation standards with smaller size and suitable for smaller delivering vehicles. The corresponding management must also use big data to improve efficiency. In the new mode of the container multimodal transportation, the big data, sensors, cloud computing, and other technical means can collect, acquire, and analyze a large amount of information in real time and realize the configuration balance in the new container through various communication terminal platforms. In this way, the high cost of equipment construction and operation can be greatly reduced and the proportion of multimodal transportation business can be increased.

Through the application of big data technology, the multimodal transportation process can be continuously optimized. In the future, it will be able to realize the same platform interconnection of key data and information of multimodal transportation box management participants, such as cargo agents, customs declaration enterprises, railways, ports, and regulatory sites so as to realize the whole process of electronic operation and real-time tracking of all links to improve the overall logistics efficiency.

7.7 Big Data Transportation and Cross-Border Logistics

In the context of globalization, the cross-border e-commerce is booming. However, due to the existence of customs supervision and other links in cross-border logistics, coupled with different national policies, legal environment, business environment, cultural environment, payment methods, tax policies and exchange rate system, the complexity of cross-border logistics has increased significantly.

In view of this, the criticality of cross-border logistics incorporating big data transportation system has become imminent.

On December 20, 2017, the “shared warehouse” cross-border ecological service platform was officially launched. This platform can provide exclusive services for the export of overseas warehouses of cross-border logistics. In the field of the cross-border logistics and customs clearance services, new service technologies such as big data, cloud services, and the Internet are used to tap idle resources such as overseas logistics warehousing and create a shared economic model of cross-border logistics.

Through the “shared warehouse” service, foreign trade enterprises can realize data interconnection and shared interaction between upstream and downstream enterprises of cross-border logistics so as to solve the logistics bottleneck faced by Chinese cross-border export enterprises.

In fact, the transportation and information flow in cross-border logistics have realized real-time networking. The cross-border logistics mode has its own characteristics, therefore the phenomenon of data fragmentation always exists. In addition, the personalization demand for cross-border logistics is increasing. Fast, efficient, safe, and personalized cross-border logistics solutions cannot be separated from the combination of big data, Internet of Things, and cloud computing.

The following are the promotion directions that the cross-border logistics can obtain through big data technology in the future:

(1) **Establishing a unified cross-border logistics big data platform.** At present, a unified cross-border logistics big data platform urgently needs to be formed. The platform is led by government departments and it develops and integrates the enterprise resources to form the fourth-party logistics services to build a smart cross-border logistics, which is an important solution to the problem.

For example, on December 26, 2017, at the China–Myanmar Border Economic and Trade Fair, the construction of Southeast Asia smart logistics big data center was officially announced, which reflects the future development direction.

On the new cross-border logistics big data platform, the whole service chain should be taken as the platform to integrate all kinds of resources on it so as to provide the cross-border logistics data information and the value-added data service chain solutions.

(2) **Establishing a cross-border logistics big data information center.** At present, China develops economic and trade exchanges with its neighboring countries and regions. In this process, the production, storage, and transportation of goods generate data on a large scale. Among them, the wide variety of data, the uncertainty of management between data and model, the diversity of logistics objects, and the more advanced tools for data collection, analysis, and processing are all features that traditional technology cannot completely cope with. Therefore, to establish a cross-border logistics big data information center, it is necessary to focus on capturing, cleaning, integrating, selecting, and updating the data in each period and link of logistics cooperation among countries and regions, and then use data mining technology to mine the potential information of data so as to realize

the value-added of data. Finally, the results should be presented in a visual way so as to achieve accurate guidance for the improvement of cross-border logistics service quality.

(3) **Building a big data trans-nation logistics system.** The construction of cross-border big data information center is the first step to realize the trans-nation logistics system of big data. The intelligent logistics system based on big data technology can make online information and offline logistics integrate with each other and realize linkage development.

First of all, we should improve the accuracy of logistics information identification in the process of cross-border logistics through the digital and visual technology in big data.

Secondly, we should make use of big data to define the logistics resources that need to be configured according to the logistics demand and conditions expected by the customers served so as to plan a more efficient one-stop solution for cross-border logistics.

Thirdly, the effective combination of big data technology, GPS system, and geographic information system (GIS) shall be made use of to accurately obtain the real-time information of the location of goods in the logistics process. Further, by using the RFID technology, the state of goods in different time points and different locations can be collected in real time. The aforementioned data and information are shared with customers synchronously so as to adjust and optimize the real-time logistics plan.

It should be noted that there are some risks in cross-border logistics. For cross-border enterprises that use their own logistics, when products enter the local market, they may no longer be popular and unsalable. In this way, it will take a huge cost for the products to be transported back to the home country, so they can only be destroyed as a result. But with big data analysis, the enterprises can effectively integrate big data, analyze global commodity demand and freight information, and judge the market that the products are launched through data analysis results. For example, big data technology can analyze the product preference of the region through the integration of customer's click, browse, preference, evaluation, and purchase data so as to guide the logistics operation.

(4) **Improving the utilization rate of overseas warehouse.** Warehousing management and utilization efficiency are important factors that affect logistics service level. Many cross-border logistics enterprises

are passive in the utilization of overseas warehouses due to the lack of data assurance. In fact, the basis of decision-making for setting up overseas warehouses depends on big data analysis. An enterprise must make a systematic analysis according to its own business scale, economic situation of trade place, logistics order, and other data, and judge whether to rent or build its own overseas warehouse through direct results. Then, according to the economic situation of the trade place, the number of consumers, the maturity of the traffic network and other data, the optimal location should be selected.

In the aspect of intelligent management of overseas warehouse, the big data technology can integrate the scattered and extensive data resources of overseas warehouse through data mining and analysis so as to build a logistics mode of unified and coordinated management of Internet and big data technology. Through data analysis, the integrated classification of goods in overseas warehouse is realized and the integration of goods from warehousing, sorting, delivery to information query is realized.

The application of big data transportation technology provides a new service mode for the cross-border logistics. With such services, the cross-border logistics can move from traditional logistics to modern logistics and ecological logistics, change the past empirical thinking to establish the data thinking and improve the core competitiveness.

7.8 Big Data Transportation and Aviation Logistics

Aviation logistics refers to the efficient and effective circulation and storage of goods from the place of supply to that of receipt, with air transportation as the main mode of transportation so as to meet the needs of customers. The advantage of aviation logistics lies in the multi-functional and integrated comprehensive service. The application of this logistics mode can organically combine the resources of transportation, storage, loading and unloading, processing, sorting, distribution and information, thus forming a complete logistics supply chain.

For this reason, the freight volume of domestic aviation logistics is growing at an average annual rate of 26.8%, and China has become the place with the greatest potential for the development of aviation logistics in the world. This situation also gives more space for the combination of aviation logistics and the big data transportation technology.

At 1:00 a.m. on January 1, 2017, Yuan Shunzhou, general manager of Shanghai International Airport Co., Ltd., announced that the "Pudong Punctuality" Airport-Collaborative Decision Making system (A-CDM) jointly developed by Shanghai Pudong International Airport and VariFlight was launched.

With the launch of this system, the operation of Pudong Airport has officially entered the era of "intelligent operation", which indicates that in the process of building an intelligent airport, Shanghai Pudong International Airport has applied big data to the production field and realized the automation, visualization, and intelligence of operation data.

Over the years, different enterprises, airports, and airlines have accumulated a large number of logistics data. However, because many valuable data are not fully utilized, most of these data systems can only do simple accounting management and lack decision-making management function. With the fierce market competition in the logistics industry, it is necessary to use the big data technology to make decision-making analysis on the data of aviation logistics, which is conducive to improving the source of goods and economic benefits.

The combination of big data and aviation logistics industry is mainly reflected in the following aspects:

(1) **Effectively managing the booking service items.** Customers' expectations to aviation logistics are first and foremost based on the ability to find the carrier information they want as quickly as possible. Aviation logistics enterprises can create personalized products and prices for their own services through the behavioral data on mobile websites, mobile applications and social media, and the correlation analysis of these data.

    Through the effective management of the booking service items, the aviation logistics companies can ensure that customers can quickly get the information they want, and also greatly improve the possibility of being booked after the introduction of information.

(2) **Improving the logistics service ability.** In fact, the aviation logistics is also one of the earliest participants in big data which have little difficulty in collecting data. But on the practical level, they must selectively grab useful data and turn them into useful information.

    For example, the aviation logistics enterprises use big data to understand customer needs, so that the customers can select the conditions such as freight departure location, destination, price, time, or guarantee priority according to their own needs, and then the system automatically recommends the optimal multimodal transportation plan. In this way, the time, route, position, historical support, and sales price can be clearly displayed on the terminal used by the customers.

(3) **Reforming the management of shipping companies.** The operation of aviation logistics enterprises is the same as that of general logistics enterprises. Therefore, the goal of big data mining and utilization of aviation logistics enterprises needs to focus on improving the level of revenue, reducing costs, and improving customer satisfaction.

For example, the revenue management of aviation logistics is the core of the pricing optimization information system for logistics enterprise, which must be optimized by a large number of data. At the same time, it is an inevitable trend in the future to take big data and cloud computing as the auxiliary operation of revenue management system to ensure that the role of revenue management system can be significantly improved.

In the long-term future, in terms of collecting and predicting weather information, scheduling information and security management, the management of aviation logistics enterprises is also inseparable from the information provided by the big data and cloud computing. It is the accurate budget made under the support of these information that can ensure the smooth operation of enterprise business.

It can be said that in the aviation logistics management, each system needs a large amount of data to support. After each system enters the big data transportation system, its performance will be improved to various degrees. In the future, the aviation logistics enterprises will integrate and establish large-scale database on the existing data processing system, process a large number of data information, improve operation efficiency, and win more customers.

7.9 Big Data Transportation and Non-Truck Operating Common Carrier

The non-truck operating common carrier is the latest rising branch of the global logistics industry. The non-truck operating common carrier does not need to have vehicles to carry goods. Such enterprises mainly build logistics information platform through Internet and other technologies to intensively integrate and scientifically dispatch logistics resources such as vehicles, stations, and goods sources. With the efficient use of the Internet, the enterprises with non-truck operating common carrier will become a new trend in the development of China's industry.

On July 20, Yunnan's first non-truck operating common carrier platform, Yunnan Intelligent Transportation, went online. At the same time, the big data transaction center of Yunnan logistics alliance was launched simultaneously on the same day.

The platform is an innovative business model of deep integration of mobile Internet, the big data technology, and the freight logistics industry. In terms of intelligent quotation and intelligent scheduling, the platform helps shippers find carriers with high-cost performance more easily, while the freight forwarding companies get more stable organization sources. The actual carrier companies or freight drivers are more secure in finding sources, collecting freight charges, and improving the utilization rate of freight vehicles.

In addition, the online platform also has real-time tracking of driver positioning, which not only provides the logistics company with the visualization data of transportation trajectory during the whole process but also improves the efficiency of logistics.[3]

The enterprises with non-truck operating common carrier not only promote the social division of labor and cooperation and effectively improve the comprehensive benefits of social logistics but also objectively promote the logistics industry in China to transform and upgrade toward intelligent, service-oriented and collaborative industries as a whole. In this process, the big data transportation system naturally plays an important role.

7.9.1 *Risk control system established and guaranteed by the big data*

In order to get rapid development of the non-truck operating common carrier industry, in addition to the monitoring and operation of the government and enterprises themselves, efforts should be made to avoid risks and ensure transaction safety. Therefore, we must rely on the big data

[3]Reported by yunnan.com on July 20, 2017, Reporter Zhao Gang.

information technology to develop, establish, and ensure the effective operation of risk control management system.

With the big data technology, the non-truck operating common carrier platform can establish a three-way risk control system. Before the event, the platform shall establish a strict qualification review and identity authentication mechanism, and each business owner and carrier shall pay the deposit in advance and enter it into the database. In the process, the non-truck operating common carrier platform acts as the monitoring role, making big data technology be responsible for the process management and supervision, call and scheduling, tracking and monitoring, risk assessment, and large freight settlement in the whole transportation process. In the post event monitoring, the non-truck operating common carrier platform also needs to connect the data with the insurance company and create the corresponding guarantee communication mechanism.

7.9.2 *Improving the user experience and stickiness*

As a new type of logistics, the non-truck operating common carrier must effectively improve the user experience in order to obtain the continuous increase of market share. Therefore, we should make full use of the big data technology, so that the platform can collect and process the massive dynamic vehicle source and cargo source information with strong multilateral nature. On this basis, we can also use the quantitative information, combined with the weather, temperature, social conditions, and other data of a region through the complex core algorithm on the platform, to obtain the demand data trend of the region in the future.

In addition, the platform can analyze the historical delivery time, delivery route, type, batch and vehicle use of the owner through big data, effectively grasp the delivery law and demand of the owner and push the source forecast and the empty vehicle information to the owner.

Correspondingly, the platform can also analyze the carrier's historical departure frequency, driving route, cargo type, and batch so as to grasp the carrier's preference and flow direction and push the source forecast and recommendation to the carrier.

Under such data analysis, the enterprise with non-truck operating common carrier will be able to greatly change the logistics experience of the shipper and the carrier. On the one hand, the cargo owner can get a fast, safe, and low price transportation capacity; on the other hand, the carrier can plan the departure and route in advance according to its own business needs and route preferences, reduce the dwell time and detour caused by the intermediate link, and save the cost of the driver.

7.9.3 *Connecting with the public platform to ensure transportation quality*

The national public supervision and service platform for road freight vehicles, the public platform for road freight vehicles, is the largest Internet of Vehicles in the world at present and also the industrial and public welfare platform for data collection, information exchange, and third-party supervision for freight vehicles. The non-truck operating common carrier can connect with this public platform to obtain data services so as to ensure the quality of transportation.

It is an important link in the business chain of enterprises with the non-truck operating common carrier that the logistics enterprises with non-truck operating common carrier shall independently monitor the transportation process. Through the big data platform, the enterprises can use hardware settings such as vehicle real-time position interface, track interface, drive in and drive out interface to connect to the public platform of road transportation vehicles of the Ministry of Transport, and timely find out possible problems in the transportation process. For example, vehicle mileage utilization, transportation efficiency, cost control, and other important issues shall be actively avoided at the first time.

In terms of capacity supplement, the carrier enterprise can also mine the vehicles entering the network by providing the vehicle demand to the public platform of road transportation vehicles, through the data indicators such as vehicle location, track analysis, vehicle portrait and, safe driving index, and the core algorithm of big data so as to obtain the nearby high-quality vehicle source.

In the future, by using the big data technology, the logistics enterprises with the non-truck operating common carrier can further deepen the integration of resources, obtain full technical support, and process services so as to make substantive progress in the market competition, and play a positive role and value for China's logistics industry.

7.10 Big Data Transportation and Logistics Costs

Like any commercial activity, the logistics also needs to spend necessary costs, including the labor, material, and financial resources pointed out by the enterprise in a series of activities such as packaging, handling, loading and unloading, transportation, storage, circulation, and processing.

In fact, the calculation of logistics cost is relatively complex. This is because many logistics costs are hidden. How to calculate logistics costs depend on the calculation scope and method. To effectively manage, control, and even reduce the logistics costs, enterprises must use the vision and methods of big data transportation.

At the New Generation Global Logistics Summit held on December 11, 2017, Wang Zhenhui, CEO of Jingdong Logistics, said that through the use of the big data technology, Jingdong has created an unmanned intelligent logistics system for the whole supply chain of warehousing, sorting, transportation, distribution, and customer service. In 2012, the inventory turnover period of Jingdong was more than 30 days. By 2017, although the daily average order of Jingdong increased 8 times, the inventory turnover period still remained at this number, far below the industry average. Through such intelligent logistics system, Jingdong will reduce the cost of rural logistics by 50–70%.

Wang Zhenhui said that with the application of big data, cloud computing, and artificial intelligence in the field of logistics, it will also bring further reduction of logistics costs so as to achieve a significant increase in the operational efficiency of enterprises.

In fact, the logistics cost is not only a problem that enterprises need to face and solve but also a field that the government and society should pay attention to. The reason why the poor and remote areas are underdeveloped or even backward lies in the fact that the logistics price they bear is far higher than that of the first-tier cities and important cities. If the logistics cost is effectively reduced and the logistics price is decreased, more employment opportunities and business chances will be brought directly so as to drive these areas out of poverty and develop economy.

So, how to use the big data transportation to effectively overcome the obstacles of natural geographical conditions and further reduce logistics costs?

7.10.1 *Effective cost saving at planning level*

We take home logistics and supermarket distribution as examples. For a long time, these urban distribution within 30 kg mainly relies on manual planning and implementation. Many enterprises have always chosen to schedule manually, including calculating how many cars, routes, manpower, and hours of working time needed. Even the experienced dispatchers do not necessarily ensure that the resulting logistics solutions are perfect. As a result, the vehicle's full load ratio is unstable and resource waste occurs from time to time.

On the basis of big data, intelligent scheduling platform can carry out precise calculation based on optimization algorithm. In the shortest time, a scientific scheduling solution can be formed to generate different distribution solutions for different distribution scenarios.

Obviously, such a formation mode of solution not only saves the manual scheduling time but also guarantees the scientificity of the distribution process and avoids waste.

7.10.2 *Optimization of logistics cost accounting method for manufacturing enterprises*

The management of logistics cost involves not only logistics enterprises but also manufacturing enterprises as customers. For the latter, its traditional accounting system is difficult to store, transform, and process a large number of logistics cost-related information, especially some hidden costs, which is more likely to cause cost loss.

On the cloud accounting platform based on big data technology, the logistics cost management model of manufacturing enterprises constructed through cloud technology can effectively collect, process, and analyze the logistics information generated in the transaction activities of manufacturing enterprises, solve the needs of data mining and process the relevant data of these logistics costs to form results to provide data support for statistical accounting.

7.10.3 *Mining potential demands of distribution customers to realize strategic extension*

The strategic significance of big data technology lies in the specialized processing and value mining of the mastered information. In the application of logistics big data technology, we can find out the potential needs of customers, such as forecasting the type and quantity of product sales, and analyzing the medium and long-term development of the market.

Due to the gradual deepening of the application of big data in the logistics industry, the data sources obtained and used by the logistics industry in the future are not only within the narrow sense of the logistics industry but also a large number of external information. Through the sharing of these data, the logistics enterprises will be able to seek customized services for every customer. Through such changes, we can reduce the cost of common logistics at the macro level.

At present, the logistics cost is an issue of concern to the whole society. Only by giving full play to the magic role of big data transportation technology and analyzing the problem can we continuously break through the original technical and management obstacles. In fact, this has become an important topic for the benefit of China and the people, which is worth every logistics person and every logistics enterprise to pay for.

Chapter 8

Big Data Transportation and Transportation Planning

The big data transportation can command vehicles to pass quickly, help travelers to find the best route and certainly help traffic managers to plan traffic routes, roads, stations, hubs, transportation capacity, etc. The big data transportation uses accumulated data for in-depth mining, reasonable planning, and ultimately making the best choice. This is much faster, more reasonable, and more labor-saving than in the past when only planning transportation by manual.

8.1 Big Data Transportation and Route Planning

The speed of urban development is accelerating. How to the use big data to plan public transportation routes is an important means to solve urban traffic congestion and improve traffic efficiency. With the cloud computing ability of the big data, the route planning is more reasonable and accurate so as to effectively improve public transportation services.

8.1.1 *Problems in traditional route planning*

For urban traffic, route planning is the foundation. Only by establishing a perfect route planning scheme can we lay a foundation for the later network line detail planning, road planning, hub planning, etc. The route is the big context of traffic. With clear and perfect planning of the big context, the network lines of small branches and nodes and hubs between them can present the purpose of multi-point dispersion and reasonable crossover.

After 2008, the rapid increase of automobile ownership in China has brought great challenges to urban traffic, of which most cities were often unaware when planning their previous traffic routes. Therefore, most of the urban traffic in China has the following problems:

(1) **No Reasonable Network.** The traffic routes in most cities have many checkpoints and blockages, and the exits are not smooth. The problem of east–west and north–south connectivity is becoming increasingly serious. Especially for the first and second-tier cities, many transit lines need to cross the city center and residential areas, which aggravates the contradiction of road traffic. The previous route planning cannot meet the current needs.
(2) **Confusing Road Function.** On the same road, the function orientation is often very confusing, including shops, hospitals, schools, shopping malls, office buildings and hotels, so the problem of traffic interference is very serious. Compared with Europe and the United States, where most of the received functions are very clear, although the number of vehicles is also large, the traffic jam problem is not as serious as that in China. The road cannot play a role, and there are obvious loopholes in route planning, which are important reasons affecting the traffic quality.

The reason why such problems occur in the traditional traffic route planning is that the previous method is based on the "four-stage" method. The "four-stage" method was born in the 1960s. It is based on a large survey of urban residents' travel, usually conducted in 5–10 years, with a low sampling rate, and often between 2% and 5%. Due to the long time and low sampling rate, the accuracy is limited, and there are obvious problems in traffic prediction.

In the era when automobiles are not widely used, the loopholes of this planning method will not be exposed. However, with the coming of the automobile era, the problems of four-stage" planning method becomes more and more obvious, which eventually leads to frequent traffic problems in the urban areas.

8.1.2 *New mode of big data transportation planning*

The continuous development of big data technology has brought new changes to urban traffic route planning. Through the use of onboard data equipment, smartphone equipment, and smart monitoring equipment, all the data are no longer isolated and start to be effectively integrated. Therefore, the large sample or even full sample data of residents' travel has been effectively captured, which can make practical judgments on the development trend of urban traffic.

The characteristic of big data lies in the data statistics through the method of computer database, which is far superior to the method of artificial statistics in terms of efficiency and accuracy so as to realize the purpose of continuous observation and analysis. The data analysis mode of traditional traffic planning is "open-loop", and too many factors are affecting the final survey data; while big data will form a "closed-loop effect", and all information will be statistically analyzed in the database, so the data will be more accurate and can promote the sustainable development of urban traffic.

Big data for traffic planning is to capture the running state of urban traffic system in real time through more abundant collection means so as to provide effective amount data for traffic planning, diagnosis, and regional coordinated control.

(1) **Internet of Vehicles technology.** The Internet of Vehicles technology realizes the information capture between vehicles and vehicles, vehicles and roads, which greatly improves the efficiency and safety of road utilization.
(2) **Data analysis.** In addition to real-time monitoring, with the help of smartphones and other devices, the big data technology will carry out continuous data analysis on the behavior habits and traffic demand of travelers so as to get the traffic demand of different groups of people, help to formulate targeted traffic demand management policies, make targeted adjustments to the road and promote the effective use of road traffic resources.

(3) **Data statistics of social platforms.** Data statistics of social platforms can also play a role in supporting traffic planning. Social platforms include WeChat, Weibo, QQ, and post bar, which are currently commonly used by the public. Through the relevant traffic evaluation, the planning department can understand the public's attitude toward traffic policies and measures so as to make corresponding regulations and reduce the adverse impact of traffic policies and measures.

8.1.3 *Construction of big data "traffic brain" in Zigong*

The big data are becoming more and more important for route planning. Therefore, many cities are also actively exploring. For example, Zigong's "traffic brain" is a flexible application of big data.

On December 25 of 2017, West China Metropolis Daily reported:

"In 2017, Zigong in Sichuan Province invested more than 50 million yuan to comprehensively upgrade the construction of Zigong's 'intelligent transportation' system. Zigong has upgraded the traffic signal control system, issued traffic guidance system and built traffic data acquisition system to solve the problems of backward functions, insufficient quantity, multi-party construction and lack of maintenance in the existing traffic signal control, traffic flow acquisition and traffic guidance. At the same time, Zigong integrates all kinds of road traffic resources, builds the 'traffic brain' of the city, digs into massive traffic data, deconstructs the operation rules and characteristics of urban traffic, achieves accurate, scientific and normal management of urban traffic and strives to make people's traffic more safe, convenient and efficient".

Through the big data traffic route planning of Zigong, we can see that the application of big data is no longer limited to a specific receiving, can cross the restrictions of the administrative region, so that the traffic planning of the whole urban area presents a unified and linked situation, and all departments can effectively combine (see Figure 8.1).

Through the integration of "data warehouse" in different areas, regions, and fields, the mode of integrated utilization of public transportation information will be brought into play, so as to have a more comprehensive understanding of urban routes. Meanwhile, combined with the comprehensive data of meteorological, urban construction, and insurance departments, as well as the data of public transportation IC card, the urban traffic distribution can be effectively obtained, which will make the route planning more accurate and effective.

Figure 8.1 2017 Zigong Dinosaur International Marathon Traffic Organization and Security Command and Dispatch.

Source: Zigong Traffic Police.

8.2 Big Data Transportation and Network Line Planning

The network line planning directly determines the congestion of urban traffic. Because of the obvious loopholes in the past traffic planning mode, the urban network line planning is unreasonable, there are too many dead end roads and unbalanced cross roads, and the urban congestion problem is becoming increasingly serious. Most cities face the problem of network line congestion, and usually adopt the methods of widening roads and increasing mileage. Although these methods can solve the problem in a short time, it cannot solve the problem of urban congestion effectively in the long term. At the same time, too many human, material, and financial resources are involved, which causes people's dissatisfaction. Some cities have repeatedly demolished and repaired a street within one year, which not only fails to effectively achieve the purpose of network line planning but also causes further congestion of network.

Therefore, the big data technology must be introduced so as to effectively improve the capacity of network line planning and further enhance the urban transportation capacity.

8.2.1 *The significance of big data technology for network line planning*

The network line planning is more detailed for the urban route, involving not only the transportation department but also the urban management department, the urban construction department, and so on. The application of big data technology lies in the effective statistics and refinement of the data of each department so as to form a perfect traffic prediction model to effectively simulate the future traffic operation state. Especially for the network line, whether there will be frequent vehicle collisions, whether there are schools around the network will be effective statistics and prediction so as to bring guidance and suggestions to the network cable planning.

(1) **Paying attention to all data.** For the network line planning, the capture and analysis of all data is the key. A single absolute number is not important. Only when it is combined with the surrounding data points, can the network line design be more effective. For example, when there is a primary school on a certain network line, the vehicle data on the primary school network line can be quickly and effectively captured, and the surrounding network line can be effectively analyzed, so that when the impact of traffic on the school is avoided, the surrounding network line can be used for drainage to realize the complementarity of traffic resources.
(2) **Accepting mixed data.** The characteristic of big data is to be able to capture a large number of data in order to formulate laws and future trend. The huge data in the traditional data statistics model is completely incomparable. Traditional models often cost a lot of money and experience, but the data can only focus on a certain detail, such as the vehicle data in front of the railway station. The data is too single, which will bring a certain degree of "deception" to the network line planning, resulting in the problem that cannot be really effectively solved.

The big data can accept mixed data, including the basic vehicle information, weather, season, and surrounding specific environment.

For example, when there is a large exhibition hall on a network line, there will be a regular surge of vehicles, so the network line design will be adjusted accordingly. Only by accepting a large number of mixed data from the macro point of view can we get the accuracy in the micro level, which is the key to the network line planning. The network line belongs to the micro field, so we should pay more attention to the capture of macro data, and the network line planning can be reasonable by combining with the surrounding information of multiple network lines.

8.2.2 *The integrity of network line planning*

The network line planning focuses on the integrity, which can effectively combine the data of streets with different attributes, improve the efficiency of network line application through the application of big data, and realize the overall operation with the seamless combination of public transportation, effectively reducing the confusion and congestion of urban traffic.

For example, for the adjustment of the public transportation network, the transportation planning department can use the big data to make effective statistics on surrounding residential communities, rail transit trunk lines, venues and facilities so as to adjust public transportation lines and demonstrate the purpose of "one road, one line and convenient transfer". This kind of network line planning can realize that after passengers enter the network line, they can quickly change to reach any place and gradually change the previous single line adjustment mode into a new mode of regional overall adjustment. The significance of big data for network line planning is to form an overall linkage mode by starting from the whole.

At present, Shanghai's network line planning is the ripest in China. With the help of the big data technology, Shanghai's public transportation has been significantly improved. In particular, the network line planning adopts a hierarchical mode to make public transportation become the main body to solve urban traffic.

The "transit metropolis" is the future transportation orientation of Shanghai. Through the application of big data information technology, Shanghai sets the congestion level for the network line and then adopts the corresponding management mode. With the real-time release of information on the Internet, the public transportation department will get relevant data at the first time. According to the actual situation, the public transportation department will adopt the mode of restricting the outside license plate of expressway, and try to restrict the outside license plate of the road in the area of frequent traffic jam to improve the utilization rate of public transportation. Shanghai will further expand its capacity. In 2015, 140 kilometers of the bus lanes were newly built in Shanghai, while the original 161.8 kilometers of bus lanes were re-evaluated and re-adjusted based on the use of the big data.

The highlight of Shanghai's network line planning is the "last kilometer". Almost all communities and streets are included and residents can quickly enter the bus station. In Taopu area, it is found that the traffic flow is too large through big data analysis. Therefore, "microcirculation pilot" is carried out. Buses less than 7 meters long are selected, which can directly enter the large communities, effectively avoiding the congestion of these road sections. At the same time, the taxi reception points and the bicycle sharing parking points at rail stations, hospitals, schools, and community service centers are further improved and radius services and network lines are well planned.

It is the key to solve the problem of urban traffic to upgrade the network line planning mode from the past single to the whole, and to integrate the network line in the form of big data. We should study what the characteristics of the network lines are in the area, which streets have too much traffic flow and which streets are suitable for emergency diversion so that the traffic operation of the whole area can be orderly.

8.3 Big Data Transportation and Road Planning

The road planning involves the details of traffic planning. The route planning and network line planning start from the macro perspective, and explore the characteristics of regional planning through the capture of big data for large flow; while planning the road, we should focus on the details to solve the actual problems of the road.

8.3.1 *Disadvantages of traditional road planning*

The traditional road planning often makes decisions through the rapid judgment of the design department based on experience and current situation. Usually, there is an experienced designer in the design institute as the leader, so it is easy to have the problem of personal will and team will in charge of operation. There are even smaller design units acting blindly and using "the copying means" to make road planning, which leads to more serious problems and completely does not fit the current needs of "Internet +" big data application.

A simple example best illustrates the problem:

An industrial park has been built in the north of a city. Every morning, vehicles will travel to the north of the city obviously, and at the same time, the number of vehicles returning from the north of the city at the end of work will increase significantly. This is the "tidal effect" in many cities. If according to the past design where the road is narrow, the variable road is insufficient, and the traffic lights cannot be intelligently adjusted according to the peak period, it is easy to cause serious traffic jams sooner or later, which can bring great trouble to the traffic managers.

Therefore, in the road planning, it is necessary to conduct preliminary research through the big data, such as using the pass-by information and traffic flow data collected by electronic police and video detector to find the rules of this section so as to make the road planning more reasonable.

For both the statistics of traffic vehicles and the environment around the road, the traditional planning has some disadvantages. For the road planning, it also needs to include the design of road space and landscape. Buildings, greenery, ground pavement, street facilities, outdoor advertising, landscape nodes, parking lots, lighting, and human activities, all belong to the surrounding environment of the road. In the past, the collection of these information often used manual observation, which led to limited data capture, simple data structure, and insufficient depth, and could not effectively provide data support for road planning and reconstruction.

8.3.2 *New ideas of road planning brought by big data*

Compared with the traditional road planning mode, the application of big data technology mainly starts from these perspectives to change the thinking of road planning:

(1) **Network data.** The network data, including hot spots and WiFi data that are widely used by residents, will record users' data including video website, Internet of Things, Weibo, blog, podcast, or forum.

These data can effectively show the hot spot map through the big data platform, and analyze the places which have the maximum flow of people. Therefore, when setting up bus stops and taxi stops, it can effectively target and maximize the use of resources.

(2) **Sensor Data.** The new traffic sensor can effectively sense the environment of the road and continuously send the data back to the platform. The new sensor can be placed on street buildings, street furniture, street greening, and street advertising, which can quickly capture the changes of the road and provide real-time data. Therefore, with this data, there are effective data references for traffic light setting and overpass on the intersection.

8.3.3 *Big data convenience of the road planning*

With the help of big data, it can effectively and scientifically improve the road planning and also bring the most direct visual experience so as to make road planning more convenient. At the same time, the public can also join in it to further improve the practicability and traffic convenience of road design.

(1) **Early stage of design.** At the early stage of road design, actively introducing the big data will make the public understand relevant information and become the information provider from the passive receiver. The traditional road design is dominated by designers and the public participation is very limited; however, through the big data platform, the design department can disclose the information through the official website and local media, and the public can provide suggestions for the design. For example, by providing information through APP, the public can log in and upload information according to the big data capture in the smartphone, which will provide reference to the design. With the help of the big data, the designers can effectively screen the meaningful information as well as the meaningless information, which ultimately affects the results of road construction.

At the same time, due to the rapid exchange of the big data, the information of electric power, water conservancy, and gas pipelines will also be included in the big data information, so that the designer can take these contents into consideration so as to avoid the repeated demolition and repair of a road due to the laying of pipelines and effectively avoid the problem of road congestion.

(2) **Later stage of design.** At the later stage of road design, the big data can also play a role. The big data can present the achievements of the road in a direct and easy way to understand, and the public can timely provide feedback. For example, through WeChat and Weibo, it can publish the visual road design system, and explain the relevant functions and features for the public to query and express their opinions. The big data will make effective statistics and classification of these information for further reference of designers and finally develop road design that meets both traffic and public needs.

(3) **Support provision for the construction of smart street.** In the future, the smart street will become the mainstream in all cities. The characteristic of the smart street is to integrate cloud computing, mobile Internet, and other technologies through smart sensing devices of the Internet of Things so that the road is no longer a simple road, but a new street with management services and intelligent services. At the same time effectively relieving traffic pressure; the attribute boundary of the road will be effectively expanded to realize active and intelligent intimate service. For example, when we are ready to park, the display board will immediately show the location of the parking lot and the number of the parking lots. This intelligent operation will greatly improve traffic efficiency and relieve traffic pressure, and big data is the core foundation of these intelligent lives. Therefore, the application of the big data for road planning will be more in-depth and fully realize the purpose of urban intelligence.

8.4 Big Data Transportation and Station Planning

The station includes all kinds of boarding points of public transportation in the city and its surrounding areas, such as the bus station, taxi stop, and parking lots on the road. The introduction of big data into the station planning can effectively complete the allocation of traffic resources, realize the utilization of resources between stations, and reduce the operation manpower of the transportation department. According to the big data, it will play a very positive role in the optimization of the main road of the ground transportation to make the ground public transportation network and the passenger flow organization plan with multiple grindstones.

8.4.1 *Application of big data in active parking system*

The problem of parking has become a universal traffic problem in the city, which not only affects the travel time but also affects the traffic efficiency because the vehicles cannot find the parking space quickly. Therefore, parking problem is a crucial problem in station planning.

With the application of big data, the parking system will be improved effectively. Among them, the active parking system has received more and more attention.

The so-called active parking system means that big data can quickly capture the information of the surrounding parking lots and push relevant contents to users. For example, when the electronic license plate has GPS positioning function, the parking guidance system will quickly compare the location of the vehicle with the information of the nearby parking lot, and push the route for the vehicle to predict the arrival time of the parking lot. In this way, the owner can choose the parking lot or reserve the parking space in advance to realize the purpose of fast parking.

At present, Shenyang in Liaoning Province has begun to explore this aspect. In a certain scale of parking lot, Shenyang has begun to implement parking guidance system, set GIS technology in the parking lot, and provide access to the city's parking information management platform. Besides, the system is integrated with the mobile terminal, and the vehicle owner can quickly query the data, access guidance to the parking, and pay the parking fees. With the combination of big data and the cloud platform, the parking lot in Shenyang is integrated into the unified management system to solve the problem of parking.

At the same time, the active parking system can also judge whether there is a problem of disorderly parking according to the parking position. In case of any violation, the relevant information will be immediately pushed to the owner of the vehicle, prompting the owner to drive the vehicle away in time. If the driver fails to leave the current area within the specified time, the information will be sent to the relevant traffic police department for taking further action. In this way, it can effectively avoid traffic jams caused by disorderly parking and effectively use the surrounding parking lot.

8.4.2 *Application of big data in bus station*

For the application of big data to the bus stations, the most basic function is the information forecast and release. At present, in cities like Hohhot and Jiaxing, the construction of "intelligent bus platform" has been carried out. The characteristic of this platform is not only to effectively classify the statistics of each bus information but also integrate GPS/Beidou Positioning System, 3G/4G traffic technology, GIS technology, and so on. When passengers arrive at the intelligent bus station, they can quickly connect to WiFi to query the operation information and station information of the vehicle, such as the number of stops before the vehicle to arrive and the estimated arrival time of the vehicle. With such accurate data, passengers can make flexible adjustments according to the actual situation, so the order of bus stations is effectively alleviated and there will be fewer people and taxis that can eventually avoid the congestion and traffic around the bus stations.

"Station Diversion" is just the basic application of big data for bus stations. It can also provide richer data and more accurate suggestions for the construction of bus stations. At present, most cities are in the process of new urban construction. With the help of big data, the traffic department can effectively collect the crowd information on each road. The population data are maximum where the need for public transportation is the greatest. With the help of the big data, the construction of bus stations will be more reasonable, which can meet the needs of most people while avoiding congested road sections.

8.4.3 *Customized public transportation: Application of big data public transportation system*

Compared with the application of bus stations, the big data have a brand-new attempt to station planning, which completely subverts the traditional mode of bus stations, that is, the customized bus.

The customized bus, a new type of public transportation mode in China in recent years, mainly serves people in the same region with the same travel time and demand. The customized bus is characterized by one person per seat, timed and fast arrival, and at the same time, it drives on a special public transportation line. The time cost is therefore easier to control. When the passengers fill in the travel information through the dedicated mobile APP, the information can be sent to the bus group for analysis, and passengers can be recruited in the customized bus platform, and the customized bus will be finally driven through the time and destination agreed.

This series of initiatives are inseparable from the application of the big data. The big data will organize the passenger information collection, customized route release, and passenger feedback information to serve specific groups of people with common goals, which will effectively optimize the road surface, convenience, and economy while reducing the use of private cars.

At present, Beijing, Chengdu, Qingdao, Tianjin, Dalian, and other cities have launched the "customized bus" service to effectively relieve the peak traffic pressure. As there is no transfer, urban road resources have been effectively solved. Beijing bus group has carried out statistics that Beijing will add 150–200 orders every day, and the attendance rate can reach more than 90%. Especially for residential areas with concentrated office workers, the customized bus is more popular, effectively improving road efficiency. This kind of station optimization can effectively reduce the number of stops of the bus, thus ensuring smooth road surface and minimum congestion.

8.5 Big Data Transportation and Hub Planning

The hub is very important for transportation. It connects different parts of the city and interchanges between cities, which is the major artery of the city's traffic, and it is directly related to the realization of the purpose of speed and efficiency.

At present, there are serious problems in the hub planning in many regions in China. For example, Xizhimen in Beijing is located at the node of the second ring road in the northwest, which is the core of urban transportation. However, due to the early construction of Xizhimen hub, the design of overpass, metro, light rail, and bus station is in some confusion, the road network is inconvenient to connect, and there are problems of confusion and congestion. Xizhimen hub is one of the road sections with the most serious traffic jams in Beijing, which does not effectively play a role as a hub. Similar problems are reflected in many cities, so the hub re-planning with the help of big data has become the current development direction.

8.5.1 *Exploration of big data in hub planning*

Generally speaking, the transportation hub often bears huge traffic pressure including bus, metro, light rail, road and even railway, airport tasks, etc. Therefore, in the preliminary survey of hub planning, the big data will summarize these contents, query the functions of administrative divisions, economic index analysis along the project, carrying capacity of bus and subway as well as the consumption potential of commercial areas.

At the exploration stage, the purpose of big data is to continuously analyze the data and find out the characteristics of the transportation hub. For example, for bus and metro, it can be effectively collected according to the times of passengers entering and leaving the station and getting on and off the train. The longer the exploration time, the more accurate the samples will be, which is conducive to the targeted construction in the future. This is a pattern that we cannot get from random sampling before.

By using the big data, it can effectively cooperate with the information of the transportation hub, optimize the cooperative operation ability of each department, and make targeted adjustment according to the principle of quick response, flexible disposal, and negotiation cooperation.

The significance of big data transport hub is to build a multi-platform and multi-channel travel mode. For example, the big data intelligent transportation hub project of Guizhou HighTech Zone, including the Baiyun station of Huancheng railway, light rail transfer station, bus station, and social parking lot, forms a variety of comprehensive transportation platforms. This will have a very positive impact on the surrounding traffic, and all traffic scheduling is done by the platform, so it solves the situation of "each department fighting for its own" and effectively solves the traffic problem.

At the same time, this kind of transportation hub often has landmark characteristics — solving the traffic problem means that there will be a "crowd effect" quickly, forming a comprehensive platform around integrating business, hotel, shopping, and talent apartment and a new urban complex project. The application of big data will continue to diverge from the field of transportation, which has far-reaching significance to the local economy and culture.

8.5.2 *Transportation hub between cities*

With the continuous development of urbanization, the communication between many cities is becoming more and more frequent, and the transportation hub is no longer limited in the city, showing a "two city interaction" situation. Compared with the planning of urban transportation hub and urban transportation hub, the application of big data is more needed so as to realize the purpose of "working in Beijing during the day and living in Tianjin at night".

Taking the "Guangfo" transportation hub as an example, we can see the role of big data planning for inter-city transportation hubs. Not only Guangzhou and Foshan rely on the application of big data but also the Pearl River Delta has formed a relatively complete hub system to ensure that the connectivity between cities is more convenient.

(1) **Metro planning.** The metro has the characteristics of being fast, efficient, and punctual. Therefore, in the construction of two-city hub, Guangdong and Foshan take metro as the most important mode of transportation through the big data analysis, and effectively connect the key business districts and schools of the two places. For example, the first phase and West extension of Guangzhou Metro Line 7 are about 32 kilometers long, with 16 stations in total, including 10 stations in Guangzhou and six stations in Foshan, which will connect University City, Guangzhou South Station, Shunde North area, Foshan new city and other important areas.

    This construction is precisely due to the analysis of the big data. Most of the metro lines in Guangzhou are located in commercial areas, office buildings and other business circles, while in Foshan, it is the residential area, which can meet the demand of "working in Guangzhou and living in Foshan". Compared with other modes of transportation, the metro plays the most important role, so it can effectively play the role of hub.

(2) **Bus planning.** Besides the metro, the bus network is also the key hub of these two cities. As of 2016, there have been 67 bus lines between Guangzhou and Foshan, realizing the coverage of cross city bus lines in Guangzhou and Foshan, forming a new traffic pattern of "Taking Guangzhou–Foshan metro line as the backbone, Guangzhou–Foshan cross city bus as the main body, and other modes of transportation (such as taxi) as the supplement". Because of the effective application of big data, these lines can access the road with the highest demand of people and strengthen the connection between the two cities.

The transportation hub function between Guangzhou and Foshan not only greatly alleviates the traffic inconvenience between the two cities, but also effectively improves the flow of people, logistics, information and capital between the two cities. The data show that by the end of 2015, the total number of passengers transported by Guangzhou–Foshan Metro had reached 240 million, which is more than 10 times of the total population of the two cities, and the talent exchange, information, and financial exchange between the two cities have improved rapidly. "One industry prospers, all industries prosper, and transportation makes all industries prosper", Huang Zhihao, Secretary of Nanhai District Committee of Foshan, commented on Guangzhou–Foshan transportation hub. On the basis of ensuring the transportation, the creation of greater economic significance has a very strong reference significance for urban agglomerations in other regions.

In the future, transportation hubs will not only be limited to highways and metros but also include high-speed rail, intercity rail, and aviation so as to form more three-dimensional hub models. Therefore, in the construction of transportation hub, we shall first introduce the big data technology, coordinate relevant data of various departments, and carry out accurate analysis, so that each mode of transportation can play its own characteristics and the hub function can be truly reflected. Just like the Xizhimen transportation hub in Beijing, in the new transformation, the big data technology is actively introduced to solve the problems of dead end road, bottleneck road, and flyover signs so as to make efforts to effectively solve the past confusion of Xizhimen.

8.6 Big Data Transportation and Capacity Planning

The transportation capacity is related to the travel needs of the public. The capacity planning is a very scientific work, especially the application of the big data, such as the areas that need the most public transportation vehicles, areas with high waiting period due to lack of vehicles, and the roads with saturated capacity — all of these need to be done through the data. Without the big data, the capacity planning can only be discussed on paper.

At present, many cities in China have the problem of imperfect transportation capacity planning, which leads to people's frequent complaints through forums, livelihood websites, etc. Some of the lines that should be added to the operation are the constant difficulties to take buses and taxis. As a result, once these areas are in the peak period, they are full of people, which poses a great threat to the road safety and leads to the waste of public transportation resources.

In fact, the lack of transportation capacity in many cities is not really due to the lack of public vehicles, but because the transportation capacity has not been maximized. An analysis by the Los Angeles Institute found that a higher level of transportation could be achieved with only 16–54% of the vehicles. At present, the problem of insufficient transportation capacity in cities is not rooted in the increase of public vehicles, but in the optimization of transportation capacity through the capture of big data, which is the key to improve urban transportation capacity.

For example, the "intelligent bus" carried out in many cities in China is the optimization and adjustment of transportation capacity with the help of big data. Relying on the positioning technology and GIS technology, bus companies can effectively grasp the operation of vehicles, which includes the stop time of the station, the number of passengers in the vehicles, the number of passengers waiting at the station, the waiting hot spot map of the city's bus, etc. Through these data, bus companies can adjust the buses in time, so that buses with idle capacity can quickly enter the road section with saturated capacity, optimize resource allocation, and effectively alleviate traffic congestion. Therefore, this kind of "intelligent bus dispatching center" is bound to be further popularized and gradually infiltrated into the third and fourth-tier cities, even the rural intercity roads.

8.6.1 *Comprehensive utilization of big data for transportation capacity*

The transportation capacity includes not only buses, but also metro, light rail, taxi, the empty loading rate of buses, taxis and so on. Many cities have such a phenomenon that the empty loading rate of buses, metros and taxis is too high in some time periods, which is also a manifestation of improper transportation capacity and a waste of traffic resources.

Once the big data can be accessed, such problems of these types will be solved. For example, in order to avoid the delay of high-speed rail and aircraft, most passengers often choose to rush to the railway station several hours in advance. Because buses and taxis fail to arrive on time on many occasion, the time spent on the road is even longer than that of high-speed rail and aircraft. When the big data technology is applied, taxi companies can timely obtain the information content pushed by the high-speed rail transportation department and airport department, and master the time of passengers' travel. Then, taxi companies can interact with the traffic information platform, analyze the traffic information of passengers' travel, and make a prediction to pick up and drop off passengers. In this way, the passengers can get the information of nearby taxis through mobile GPS positioning, and taxi drivers can choose the best path to reach the railway station. In this way, the passengers' time is saved and the taxi is effectively used as well.

This mode has been well applied on platforms such as Didi Taxi and it is the first choice for most people. In the future, the operation optimization based on the big data will be further deepened, for example, the bus system will also be included, which can judge the road condition flexibly, allocate the idle vehicles to the key road sections, improve the operation capacity, and reduce the cost. Through the sensor, the traffic information will be timely informed to the driver, the data can be effectively captured and used, and the traffic operation will be more reasonable. Upgrading the transportation capacity planning without increasing the number of motor vehicles is the core of the future of urban traffic development.

In fact, the optimization and promotion of transportation capacity by big data has not only been carried out in the urban areas but also in the vast rural areas. Yuyao, Zhejiang Province, made an effective attempt. In 2012, the integration of urban and rural passenger transportation was reformed and the bus lines were re-planned, but there were also many problems. For example, part of the bus lines carry less than 50% passengers every day, and even there is a problem that the passenger flow of bus No. 612 is zero.

In 2016, in order to solve the problem of obvious loopholes in the bus capacity, Yuyao transportation department adjusted the bus capacity with the help of the big data, and made a thorough investigation on the actual operation of 105 bus lines in the city. After big data analysis and suggestions, Yuyao's bus capacity was realigned. Originally, bus No. 632 had two operating vehicles and 10 shifts, then it reduced to one, and the departure time was optimized. As a result, bus No. 632, which had a capacity of less than 30%, did not affect residents' travel, but significantly improved the capacity. For mountainous areas, Yuyao concentrates the departure time and frequency on the morning and evening peak, which is convenient for farmers in mountainous areas to travel. Through such adjustment, Yuyao bus company saved more than 1 million yuan of operation cost in a year, but its transportation capacity has been significantly improved. Each bus has its own research report of "work and rest arrangement".

Thus, for capacity planning, the big data can be widely used. The continuous intelligent bus transformation is the adjustment and optimization of transportation capacity planning. Using the IC card to carry out big data mobile phone and adjust the transportation capacity according to different roads and different periods will make the urban traffic operation more smooth, reduce the operation cost, and protect the urban environment construction.

8.7 Big Data Transportation and Emergency Planning

In 2003, a citizen in Nanjing opened the APP of "intelligent travel in Nanjing" before going to work. The APP pushed a message for him: there was an accident on the road the he needs to pass every day, so it is suggested to detour. Through this warning, the citizen chose a new route and arrived at his workplace on time.

Such scenes are not from a science fiction, but a reality. With the help of the big data, emergency events can be identified on time and corresponding route suggestions can be made, which becomes another key direction of the big data for traffic application.

In any country, there will inevitably be accidents or other emergency events. How to establish a perfect emergency planning scheme is one of the important assessment methods for the public transportation management. In the past, the emergency planning often had the characteristics of being temporary and sudden, which cannot effectively transmit to the public. As a result, it is prone to frictions and contradictions, and the public does not understand the governance. However, in the Internet era, with the help of the fast big data, the emergency planning should be more mature, forward-looking, and communicative, the responsibility of the future transportation department.

8.7.1 *Construction of traffic emergency system*

In the era of big data, the establishment of emergency system is no longer limited to the public transportation department, but it should take the transportation department as the core, and at the same time establish a seamless connection with the fire department, public security department, and electric power department through the network system, so that the big data can play a real role, and effectively integrate all kinds of information to achieve the rapid development and release of emergency data.

The establishment of emergency command service platform has become the key to big data emergency planning. Such a platform will play three important roles: beforehand supervision and prediction, interim emergency response and scheduling, and afterwards summary and filing.

(1) **Beforehand supervision and prediction.** The traffic command center conducts real-time inspection on each road section through the system, conducts more self-monitoring on the congested road section, and timely deals with the hidden dangers found. The data platform will share data with other departments, such as house decoration and fire drill in a certain section. Forecast information shall be sent to the public through the platform in advance to remind the public to detour.
(2) **Interim emergency response and scheduling.** In case of traffic accident, the traffic emergency platform shall immediately start the emergency plan, timely transmit the scene of the accident to the traffic command center, and at the same time, the information can also be transmitted to the hospital and the public security department at its earliest to do a good job in emergency treatment and avoiding traffic jam.
(3) **Afterwards summary and filing.** The emergency platform can effectively record the intercommunication information between various departments and save the picture. After the accident is handled, the data can be classified and consolidated to provide basis for the handling of similar events in the future.

All of the aforementioned work needs the big data support, which includes the urban road state judgment, information query, information transmission of various departments, and so on. Only in this way can the big data play a role, handle events efficiently and timely, and ensure the smooth flow of roads. Otherwise, when the accident occurs, if the traffic department does not receive the information or the medical institutions do not receive the treatment call, it will inevitably cause further traffic chaos and greatly affect the travel.

8.7.2 *Application of big data intelligent equipment*

The establishment of emergency planning is inseparable from the support of all kinds of equipment. Especially for the application of big data intelligent equipment, it will effectively improve the emergency capacity and change the scientific nature of emergency planning from the scientific and technological level. For example, the application of sensor technology, the communication technology and GIS technology will make the process of data collection faster and form the effect of "synchronous on collection, analysis and suggestion".

The intelligent signal light is a typical application. The intelligent signal light will automatically capture the adjustment cycle of road conditions, timely change the signal in case of any problem, shoot and capture the case on site, and make the plan of bus priority to alleviate the traffic problems in emergency time.

Compared with the traditional signal light, the intelligent signal light is equipped with a vehicle inspection board, which can effectively collect the data of surrounding vehicles including traffic flow, occupancy, speed, queue length, etc. Once there is an emergency, it will be adjusted according to the actual situation of the traffic flow. In this way, even in case of emergency, other vehicles can drive normally according to the indication of the signal light so as to pass orderly and keep the road smooth.

8.7.3 *Operation mode of emergency mechanism*

For the traffic emergency, the big data must be real time so as to solve the problem effectively. Therefore, the big data platform must be stable. Once an accident occurs in a street, it can effectively process information and ensure that the traffic will not be blocked.

At the same time, the installation of big data intelligent equipment should be popularized in public transportation vehicles as much as possible. Especially for buses, a comprehensive upgrade plan should be put in place to complete the installation and debugging of the equipment over a period of time, so as to truly achieve data capture. Otherwise, the data cannot form consistency and continuity, which is not conducive to the later work. Although big data intelligent equipment will generate certain financial expenditure, and later maintenance also needs to consume certain human and financial resources, but in the long run, its economic benefits and social influence are undoubtedly huge.

Especially for prediction, the big data emergency system should play an important role. For example, for large-scale activities and special service tasks, the big data should collect surrounding information at the first time, issue advance notice for possible congested road sections, remind citizens to detour, make effective statistics for surrounding hotel check-in, catering reservation and other information, find out the time period of possible traffic congestion, estimate the flow of people, and make targeted plans. At the same time, it is also necessary to cooperate with other departments to prepare for emergency support articles and unify linkage so as to truly play the role of big data prediction and prepare for possible emergency events in advance.

Chapter 9

Big Data Transportation and Major Activities

From the Olympic Games to the G20 summit, in recent years, China has been hosting more and more major world events, which puts forward higher requirements for traffic emergency response and command. With the participation of the big data, the traffic planning in major activities can be followed. Even in case of emergency, the big data can help traffic managers to predict in advance and reasonably solve the situation.

9.1 Big Data Transportation and Beijing Olympic Games

In 2008, the Olympic flame came to Beijing, which was the most important global event in Beijing. The Beijing Olympic Games, which attracted the attention of the whole world, had received 6.52 million domestic and foreign tourists and 382,000 inbound tourists during the Olympic Games, which was undoubtedly a great challenge to Beijing's traffic and a comprehensive display of the image of the capital.

So, how did the big data transportation serve for the 2008 Beijing Olympic Games? What achievements are worthy of learning and reference for future large-scale activities?

9.1.1 *Equipment application of big data*

For the Olympic Games, taxis are the most important means of transportation. The big data application of the Beijing Olympic Games to taxis is also China's first comprehensive attempt. Prior to 2008, all the expressways within Beijing's Fifth Ring Road basically completed the installation of RBMS flow meter, which can effectively capture the traffic conditions of the expressways.

The biggest bright spot comes from the taxis. When the 2008 Olympic Games began, all taxis in Beijing had been equipped with the GPS. The traffic management department could effectively locate each taxi, track it in time, and efficiently solve the problem.

Besides, the electronic charging mode had also begun to operate among Olympic enterprises. The in-depth big data cooperation between Beijing Olympic Organizing Committee and Beijing transportation department with China Mobile, China Unicom, and China Telecom could effectively identify vehicle license plate to solve the traffic accidents quickly, and store and backup these data on time.

At the same time, during the Beijing Olympic Games, hundreds of traffic incident detectors were distributed on the streets of Beijing. These devices could detect traffic accidents, road ponding, and other problems at its earliest and set their own alarms. The information was then immediately displayed in the command center, and the personnel in the transportation department handled the situation in the quickest possible time. This automatic alarm device shortens the handling of traffic accidents by 3–5 minutes, which greatly improves the rapid handling capacity of traffic accidents and ensures the effective operation of urban traffic.

The intelligent signal light were installed in the Beijing Olympic Games. Because the road network in Beijing is characterized by the mixture of pedestrians, motor vehicles, and non-motor vehicles, the transportation department in Beijing would upgrade the signal lights in an all-round way according to the current situation. The signal lights can be effectively adjusted by computers and would change with the traffic peaks and flat peaks of the roads. In this way, the traffic at the intersection optimized and the comprehensive traffic capacity of the road network increased by 15%. The signal lights in the Olympic Center also added the voice prompt function for the blind, greatly improving the characteristics of human-oriented service and further ensuring and improving traffic safety.

In 2008, the word “big data” had not been understood by the public. But the Beijing Olympic Games made a successful attempt with the help of these means. Therefore, Beijing Olympic Games can be called as “one of the most successful Olympic Games in history”, to which the Beijing transportation has made great contributions. Compared with the frequent traffic accidents in Rio 2016 Olympic Games, Beijing Olympic Games had achieved great guarantee in traffic, among which the use of these big data intelligent equipment provided accurate data for the traffic management department, carried out effectively early warning of traffic accidents, and realized rapid solution.

9.1.2 *Advance response to big data transportation*

In fact, Beijing has already started the application of big data transportation. It is because of a large number of tests at an early stage that the big data can play an effective role during the Olympic Games.

(1) **Intelligent bus.** Before the start of the Olympic Games, Yihualu Group and Beijing Bus Group had undertaken the national science and technology research plan of “intelligent operation and dispatching system of Olympic bus”, which had been officially put into operation before the Olympic Games.
(2) **Monitoring system.** During the Beijing Olympic Games, the Internet big data application mainly focused on TV monitoring, traffic signal control, and other modes. Before the start of the Beijing Olympic Games, the Beijing transportation department had formulated a complete plan and carried out intelligent dispatching. The Olympic traffic command center, Yangshanqiao traffic service command center and 38 venue group traffic command posts constituted a complete data monitoring network to monitor the road traffic in real time so as to realize the orderly parallel and harmonious operation of the traffic.
(3) **High definition digital detection system.** Before the official start of the Olympic Games, high-definition digital monitoring system were installed in the central areas of the Olympic Games. This system could carry on the omni-directional detection to the vehicle, realize the video monitoring, the vehicle recognition, the illegal detection, the flow statistics, and many other functions in the central area. The use of this system could effectively ensure the orderly development of

traffic in the Olympic central areas, and it was also the first time for China to adopt a single comprehensive detection equipment, which is more integrated and statistically more complete than the traditional multiple equipment.

9.1.3 *Application of big data system*

In addition to the big data hardware and big data response in advance, the system construction of big data guaranteed the smooth execution of the Olympic Games. Among them, the following contents constitute the big data transportation system of Beijing Olympic Games:

(1) **Mobile handling plan.** In order to serve the Beijing Olympic Games, Beijing transportation department had set up a complete mobile processing scheme. Once there is a traffic problem, the big data platform would be launched as soon as possible, and the command vehicles equipped with big data systems such as satellite communication, wireless transmission, and image acquisition would rush to the scene as soon as possible. After arriving at the scene, the command vehicles would immediately send data to the surrounding intersections through the signal system. In 2008, WeChat, and Weibo have not yet been born. Beijing transportation department prompted drivers to bypass through information board. Although the means were relatively backward, it effectively alleviated the traffic congestion at the incident location.

(2) **Monitoring system plan.** At the beginning of the Beijing Olympic Games, all expressways, main roads, and Olympic special lines had installed tens of thousands of detection coils. This is the nerve end of urban traffic management, which can work 24 hours without interruption to analyze the data of road traffic flow. At the same time, they would automatically detect the vehicles. Once it was found that there were vehicles in violation of the "odd-and-even license plate rule" restrictions; it could take photos and record them in the first instance and upload them to the data platform, effectively ensuring that the citizens abide by the laws and regulations and ensure the smooth operation of the traffic.

(3) **Bus priority control system.** For the travel, Beijing suggested that the public chose the bus system first, therefore, the Beijing transportation department had also optimized the bus services. In Beijing, a total

of 126 signal lights with priority control were installed. When the bus passes these intersections, the signal light will detect the bus and transmit the relevant information to the computer. According to the actual situation of the intersection, the computer can adjust the traffic flow, shorten the release signal event, or extend the green light event so that the delay event of the bus at the intersection can be effectively compressed to avoid the problem of road delays.

(4) **Digital traffic law enforcement system.** In order to deal with traffic violations in Beijing, 1,100 sets of electronic police were installed on the road in the past connected to the Internet, which could automatically detect nine kinds of traffic violations, including red-light running, overspeed, and so on. These information will be uploaded to the database as soon as possible, and quickly compared to lock the perpetrator promptly. At the same time, the mobile patrol car is also equipped with relevant equipment, and can realize wireless transmission, which can effectively discover the fake-licensed car. From the recognition to system terminal alarm, it can monitor 2,200 cars per hour in less than one second, achieving real high speed and high efficiency.

(5) **Expressway traffic control system.** Beijing's Ring Road, or expressway, is the main artery of Beijing's traffic. During the Olympic Games, it needs to undertake the traffic driving in the urban area on the one hand, and it is also the composition of the Olympic special line on the other. Therefore, Beijing's transportation department had established an intelligent traffic control system for the expressway, which used the intelligent signal lights of the Second Ring Road, the Third Ring Road, the Fourth Ring Road, and the main entrance and exit to capture the changes of the expressway traffic flow, and realize the function of automatically closing and opening of the entrance. When there is a traffic jam at the main intersection of the expressway, the exit signal light will control the traffic flow of the upstream auxiliary road, so that the vehicles can quickly drive out of the expressway. At the same time, the roadside information board will update the information on time to remind the drivers to choose their route so as to effectively avoid traffic accidents at the entrance and exit.

Although the term "big data transportation" was not formally born during the Beijing Olympic Games, it can be seen that with a series of big data equipment and systems, the Beijing transportation department, in conjunction with other departments, had carried out a comprehensive guarantee of transportation, and had gained rich experience for various large-scale events held in China in the future. It can be said that the Beijing Olympic Games is the beginning of the big data transportation. With the continuous upgrading of the Internet, especially the full explosion of the mobile Internet, the big data transportation has increasingly become a key part of large-scale events, providing an all-round guarantee for the effective conduct of the event.

9.2 Big Data Transportation and Shanghai Expo

In 2010, the Shanghai World Expo was another major activity held by China after the Beijing Olympic Games, which also attracted the attention of the whole world. Compared with the Beijing Olympic Games, the Expo had a much wider reach, no longer limited to sports, covering the culture of most countries in the world, so the numbers of visitors were more. According to the data, 46 countries and international organizations participated in the Shanghai World Expo, receiving 73.084 million visitors, an average of 397,000 visitors a day, which set a record in the history of the World Expo.

Compared with the 2008 Beijing Olympic Games, the big data had experienced two years of development, and its technical capacity was more mature. Therefore, the Shanghai World Expo had further improved the application of big data, especially in forecasting and other aspects. The big data played a crucial role and laid a foundation for better service of the traffic during the Expo.

9.2.1 *Big data forecast*

Before the official start of the Shanghai World Expo, the Shanghai transportation big data had officially started to operate. Through the cooperation with the ticketing platform, hotels, the catering, and other institutions had gradually mastered the relevant data during the Shanghai World Expo. Most commercial institutions in Shanghai were connected to the big data system, and the data could be counted at the earliest possible moment.

At the same time, the sales data of automobile 4S shop also entered the big data system, which improved the relevant information. With the help of this big data model, the Shanghai transportation department also made a prediction on the traffic during the Expo: in 2010, the traffic volume of Shanghai would reach 50 million person times/day, and the car ownership would reach 1.6 million. During the peak period, rail transit lines such as Lines 8 and 6 would face great challenges, especially during the World Expo, the traffic pressure from September to October would further increase.

Through big data, Shanghai transportation department had a preliminary judgment on the traffic operation during the Expo. Therefore, according to the data, relevant departments had formulated the following traffic suggestions:

(1) **Advocating to travel by bus.** In order to deal with the road traffic pressure, before the opening of the World Expo, Shanghai had been making continuous plea to encourage tourists to use public transportation at that time, especially buses. The transportation department hoped that more than 90% of tourists can visit by rail and ground bus.
(2) **Balanced entry into the park.** With the help of the big data analysis, we can see that the passenger flow in the World Expo were maximum on weekends and holidays. Due to the high reservation rate of hotels in these periods, we had formulated the strategy of "balanced entry into the park" in advance, guided the passenger flow from the peak day to the ordinary day, and set a detailed plan. Tourists were encouraged to travel from the peak time to the flat peak time and enter the park from Puxi to Pudong, so as to relieve the traffic pressure and tourists all over the world could really feel the charm of Shanghai World Expo.
(3) **Appropriate management measures.** In order to reduce the impact of the Expo traffic and daily traffic, the Shanghai transportation department had also issued other control policies, including vehicle use management and construction area management to avoid the traffic pressure not being controlled.

It is through the application of the big data that Shanghai transportation department finds out the possible problems and formulates these three traffic management plans. These three plans have set the tone of the traffic management for the Expo, and they are further optimized the by big data to make the work more detailed.

9.2.2 *Utilization of big data intelligent system*

In order to better serve the World Expo, the Shanghai traffic management department had set up the intelligent transportation integrated information platform based on the big data and the traffic information sub-platform of the World Expo Park to provide intelligent services. The big data platform of the Expo was characterized by strong data processing capability, not only for internal use but also for all tourists. When the big data encountered traffic problems like obvious road congestion ahead, it would immediately send relevant information to the Expo traffic guide, Expo traffic website, traffic service hotline, radio and television stations, variable information signs, and mobile phones vehicle navigation terminals to constantly remind tourists and provide services such as transfer scheme query. At the same time, this system is also associated with the traffic commander on the front line. The traffic commander could also obtain the traffic information in the Expo Park anytime and anywhere, so that the traffic guidance work could be started as soon as possible. Compared with the Beijing Olympic Games, the big data platform of the World Expo had been significantly improved. Everyone could quickly query relevant information through the Internet, without having to wait for the reminder from information signs.

Around the big data platform, Shanghai World Expo brought big data intelligent transportation to the extreme. There are four highlighted spots of these kinds, which provide a very good reference for future large-scale activities.

(1) **Comprehensive travel information service system.** Shanghai World Expo launched a comprehensive travel information service system. The service object of this system was the vast number of tourists. It covered all the comprehensive information of traffic modes, highly centralized the information, and could check all the traffic information conditions through a platform, avoiding the problems that need to be repeatedly inquired in the past, greatly improving the efficiency of tourists' travel and the road traffic. In particular, the service mode of "one website, one phone" enabled the tourists to get comprehensive traffic information with one click. Visitors could get the traffic information of Shanghai through many Expo traffic websites, mobile communication services, Expo hotline, special channels for traffic radio, etc., which not only brought convenience to the tourists but also to the

traffic commanders at the front line. They could quickly understand the information of the relevant road sections so as to make accurate adjustment to the traffic.

(2) **Electronic toll collection system.** In Shanghai, most road sections are toll roads. In the past, manual toll roads were inefficient, which could easily lead to traffic queues and directly affect road driving. Before the World Expo, all these sections were upgraded to electronic toll collection system, which was more flexible and diverse. It could not only realize rapid electronic toll collection but also combine a variety of preferential schemes, so it was welcomed by car owners. With the help of electronic toll collection system, the traffic resources have been effectively balanced and the speed of vehicles has been significantly improved. At the same time, the cost of transportation operation companies has also been reduced to provide better service.

(3) **Traffic monitoring system.** Similar to the Beijing Olympic Games, the Shanghai World Expo also launched a big data traffic monitoring system. After two years of technical improvement, the big data traffic monitoring system of the World Expo was more mature. It could establish a video image monitoring system for the first time to monitor the traffic flow, surrounding traffic flow, pedestrian flow, traffic security, and other aspects of the traffic intersection. All monitoring data would be sent to the monitoring center as soon as possible, covering the whole of Shanghai. At the same time, the monitoring system can give an emergency telephone alarm and dynamically issue a warning so it had a better guiding significance for the vehicles on the road. The vehicles could know the traffic emergency events ahead in advance so as to change the way and avoid obvious congestion.

(4) **Decision support system.** During the World Expo, Shanghai transportation department launched an innovative "traffic decision system", which relied on the big data to solve the problems of flow detection, dredging, and decision-making. The traditional manual mode had been completely eliminated, and relevant work would be completed in the system. The system could create models to quickly find solutions, which was much more efficient than the traditional manual mode. Because there were tens of thousands of traffic data models in the system, it could deal with almost all traffic emergencies and made accurate and scientific planning suggestions.

The four systems formed the basis of the big data transportation of Shanghai World Expo. The spatial scope of the four systems included the Expo Park, surrounding traffic control area, Expo express channel, comprehensive transportation transfer hub, rail transit network, Expo bus special line channel, Expo Park cruise channel, and other directly related areas. Moreover, these big data technologies have broken through geographical limitations, not only serving the Shanghai region but also affecting the Yangtze River Delta region, including all the space related to the World Expo. Therefore, despite the number of visitors to the Shanghai World Expo had set a record, there was no obvious omission in the orderly transportation, which provided a positive guarantee for the holding of the World Expo.

With the closing of Shanghai World Expo, these big data platform systems continue to serve Shanghai and make further development. License plate recognition data, land-use remote sensing image data, mobile communication data, and so on, have now been fully implemented in Shanghai. It can be seen that the application of big data can guarantee major activities, serve urban construction more, form a broader big data transportation system, and ensure the joint promotion of urban transportation.

9.3 Big Data Transportation and Hangzhou G20 Summit

From September 4–5, 2016, the G20 summit was held in Hangzhou, China. Compared with the Olympics and the World Expo, the G20 Summit Forum on international economic cooperation, where the heads of 20 important countries in the world gathered in Hangzhou, was more significant in politics and economy, and it was another important exhibition of China's national strength and economy. As an important bellwether of national economic development, the transportation had also become the focus of the world.

In 2016, the development of the Internet had entered a new stage, which was different from 2008 and 2010. The concept of big data had been widely used in the business field. More and more professional big data service companies had been born. Therefore, compared with Beijing Olympic Games and Shanghai World Expo, the big data transportation of G20 Summit was more mature and presented the characteristics of fine differentiation. Whether it is the transportation department, the traffic police department, or even the Zhejiang provincial government, they all had made big data transportation a priority.

9.3.1 *Attention on big data transportation from government level*

In May 2016, on the eve of the G20 Summit in Hangzhou, Li Qiang, Governor of Zhejiang Province, inspected the preparations for the G20 Summit, with a special emphasis on traffic issues. He pointed out that it was necessary to speed up the construction of a modern integrated transportation system with the interconnection so as to support the future development of Hangzhou, which was a clear guidance for big data transportation from the government level and was conducive to the rapid establishment of big data platform.

Later, Li Qiang paid a visit to the big data transportation, especially to the traffic security integrated information command center of the G20 Summit in Hangzhou. The traffic security integrated information command center of the G20 Summit had fully realized the access of big data, the integration of major transportation modes, and the real-time monitoring and scheduling of all traffic. Highway, railway, civil aviation, and other information were also included in the platform. This set of platform was the core to ensure the transportation of G20 Summit and played an important role in information collection and information transmission.

9.3.2 *Support of big data for road traffic*

At the G20 Summit in Hangzhou, for the application of the big data transportation, professional big data companies had been introduced, among which the Dt Dream was involved in the work of the traffic police detachment. Dt Dream provided a perfect traffic big data platform for Hangzhou traffic police detachment, which could analyze the traffic dynamics in real time and carry out information research and judgment at the same time. It could quickly formulate traffic measures according to the regional traffic characteristics, greatly ensuring the traffic smoothness and safety in Hangzhou.

At the same time, the system was also connected with the public transportation system, and the data of Hangzhou T-union could enter the platform at the quickest possible time. In this way, the traffic police department could control the departure shift of the bus through real-time analysis, reduce the empty car rate, and ensure a smooth traffic flow on the road; it could also optimize the route according to the peak condition of the road. In order to cooperate with the application of big data, all buses in Hangzhou were equipped with GPS positioning system, connected to the Internet of vehicles, road network monitoring, etc. Therefore, although the number of traffic polices in Hangzhou had not increased, through big data technology, the traffic situation in Hangzhou could be well known, and the route adjustment could be carried out at any time according to the current situation to avoid urban congestion.

It could be said that this system had played a crucial role in the traffic optimization of the G20 Summit. Citizens could enjoy more accurate services and shorten the travel time. Public transportation could also adjust routes in real time to achieve the purpose of "accurate and flexible scheduling" and maintain the stability of road traffic.

9.3.3 *Accurate prediction of big data*

Due to the rapid rise of the third-party big data companies more accurate prediction for the services of G20 Summit with more professional operations were possible. Among them, a series of data releases of Didi not only provided a very effective data reference for Hangzhou traffic management department but also let the citizens directly understand the possible traffic problems during the G20 period, so they could plan the route in advance to avoid traffic emergencies.

On July 27, 2016, Didi released the Big Data Report on Intelligent Travel: Hangzhou, which clearly stated that it was expected the total travel volume in Hangzhou would decrease by 22% during the G20, the average speed would increase by 45%, and the travel demand of Hangzhou people during the Summit would increase by 80% compared to the weekend. Further, due to the rapid speed of infrastructure construction in Hangzhou in recent years, the morning and evening peak of Hangzhou working day were alleviated to some extent.

In this report, Didi even made big data accurate prediction on the speed: due to the traffic restriction measures, the traffic in Hangzhou would be smoother, and the average speed would change from 24.9 km/h before the Summit to 36.2 km/h, an increase of 45%. Since most of the citizens in Hangzhou were given "seven days of summit exclusive leave", the big data had also begun to conduct vertical constraint analysis and recommended "peripheral tours" to Hangzhou citizens. In this way, the traffic in Hangzhou could be further guaranteed. According to this big data, the tourism bureaus of the surrounding provinces and cities also formulated various tourism preferential activities for Hangzhou to attract Hangzhou citizens.

This kind of accurate prediction with the help of big data was not only aimed at the beginning of the G20 Summit but also extended to the post G20 Summit. Jiang Jun, Director of the market promotion department of Hangzhou Tourism Committee, said, "according to the judgment of big data, there will be a surge of tourists in Hangzhou after the G20 Summit, which is shown by the big data statistics of relevant hotel and catering industries". With the G20 summit being held, Hangzhou had added many new attractions, and the end of the G20 coincided with the Mid-Autumn Festival and national day. Therefore, in view of the transportation after G20, Hangzhou has once again made targeted strategic adjustments to welcome a large number of tourists. The data showed that the number of orders of peripheral tourists taking Hangzhou as their destination on the weekend of September 10–11 had more than doubled year-on-year, and the proportion of tourists taking high-speed rail and long-distance bus ranked second and third, respectively, which was consistent with the conclusion of big data. Thanks to the support of big data, Hangzhou's traffic is always in good condition before, during and after the G20 Summit, and the situation of "chaos after large-scale activities" in the past did not appear.

9.3.4 *Rooting and sprouting of intelligent big data*

With the blowout of the big data technology, during the G20 in Hangzhou, the application of intelligent big data was more extensive, and more and more big data technology and intelligent product enterprises had joined in this activity, further ensuring the smooth holding of the G20 Summit in Hangzhou.

(1) **Application of Hisense intelligent bus system.** Prior to the G20 Summit, Hangzhou signed an agreement with Hisense on providing intelligent upgrades for all buses in Hangzhou. Hisense renovated 4,700 buses in Hangzhou, and for the first time in China, and it launched the function of "return prediction". The feature of this

function lay in the effective combination of the data of each bus, the judgment of the road conditions, and the accurate prediction of the return time of each bus. In this way, the bus company could adjust the route effectively, such as increasing or decreasing the shift and adjusting the departure time. This system greatly reduced the cost of manual dispatching, and made the automatic departure rate of bus system reach 90%, breaking the original record of the industry.

At the same time, the intelligent bus system also launched the bus App, which enabled citizens to carry out travel planning and obtain arrival prediction and arrival bell ring and other services at home, greatly improving the convenience of bus routes.

(2) **Establishment of transportation community.** Community culture is a new term of Internet that has received more and more attention in recent years. With the help of big data, different groups of people will be gathered together, so that the transmission of information will be more efficient and accurate. At the G20 Summit in Hangzhou, this kind of "transportation community" achieved an effective attempt. During the G20 Summit, the Transport Bureau launched the activity of "WeChat–Weibo, one end, one line, one network, one room and two groups". Through the eight community platforms of Hangzhou traffic Weibo, WeChat, APP client end, 12328 hotline, Hangzhou traffic information network, news live room, WeChat group, and QQ group, it continuously disseminated various traffic information to the public, such as passenger flow forecast, traffic meteorology, and venue ticket sales. Therefore, citizens and foreign tourists could get relevant information in the first instance, and make travel arrangements in advance to avoid frequent travel due to information negligence, which could affect the normal traffic in the city.

(3) **Application of Internet + traffic.** The rapid popularization of mobile phone, the continuous speed up of 4G network and the increase of WiFi convenience make the application of big data more popular among the people. The "Internet + traffic" mode has become the mainstream. This mode was widely used during the G20 Summit. When passengers travel, no matter bus, metro, or shared bicycle, they can scan the code through mobile phone, and they do not need to apply for additional bus card so as to realize "one machine in hand, free travel in the world". Convenient ticket purchase and payment methods greatly optimized the transportation, so the traffic efficiency

of Hangzhou had been significantly improved, and the traffic jam on the road had been significantly decreased.

From the Beijing Olympic Games to the Shanghai World Expo, and then to the G20 Summit in Hangzhou, we can see that the big data are more and more widely used in the field of transportation, and at the same time, it has increasingly infiltrated into the people. Every citizen has become an important part of the big data transportation. Only when such a situation is formed, can urban traffic be further improved, and the role of big data become more and more obvious, forming a "border effect", breaking through the simple traffic field, and realizing cross-border cooperation with tourism, economy and culture, which is more obvious help for urban construction.

9.4 Big Data Transportation and One-Belt-One-Road International Forum

On May 14–15, 2017, one of the "One-Belt-One-Road" International Cooperation Summit Forum was held in Beijing, which was an important diplomatic event hosted by China and it was of great significance to promote international and regional cooperation. As the initiator and organizer of One-Belt-One-Road Forum, China must ensure the smooth convening of the forum. At the same time, this was also the most important international activity after the Beijing Olympic Games.

This forum also attracted many heads of countries, government officials, and institutions from all over the world and 29 foreign countries and government heads and three heads of important international organizations: the Secretary General of the United Nations, and the Chairman of the International Committee of the Red Cross attended the summit forum. About 1,500 distinguished guests from all walks of life from more than 130 countries attended the forum as official representatives. More than 4,000 journalists from the world registered to report on the forum. It could be seen that the forum was of high specifications and had far-reaching influence.

For this important international forum, the Beijing transportation faced another challenge after the 2008 Olympic Games. For the "One Belt, One Road", what role did big data transportation play?

In recent years, with the continuous development of big data technology, the traffic in Beijing has formed a relatively perfect big data transportation mode. Therefore, for the One-Belt-One-Road International Forum, these new technologies have been widely applied. These technologies not only served the One-Belt-One-Road International Forum but also set up a perfect big data transportation network for Beijing, while ensuring the smooth convening of the international forum, they also brought real convenience to the Beijing people and greatly improved the traffic condition of Beijing.

9.4.1 *Traffic restriction measures under big data*

It is an international practice that major conferences should be regulated and traffic restricted. The same was true for the One-Belt-One-Road International Forum. In the past, the traffic restriction was often judged by experience and had a delay in execution, which would bring great trouble to the normal travel of the public. But for the One-Belt-One-Road International Forum, the Beijing traffic department used big data to carry out a long time of information collection for key sections. With the help of intelligent traffic lights, road monitoring equipment, and other large data channels, the relevant information was released in early May 2017, so that the public could prepare ahead of schedule.

At the same time, the restricted routes were also optimized according to the big data judgment. Instead of being blocked by most roads in the past, they only restricted the roads that had an impact on the forum. This not only ensured the smooth development of the forum but also ensured that the daily life and work of the public were not affected. At the same time, due to the active application of big data, Beijing traffic management department also issued relevant bypass suggestions, which further guaranteed the normal travel of residents.

(1) **Traffic control measures were taken for roads around Yanqi Lake International Metropolis in Huairou District.** From 0:00 on May 7 to the end of the activity on May 15, from the intersection of Fanqi Road, Huidu Road, Huairou District (not included), via Fanqi Road, Yanqi Lake West Road to Baiyashanzhuang Bridge (not included), Yanqi Lake Road, and Yanqi Lake South Road, except for vehicles with special certificates of "Summit Forum", other vehicles were prohibited to pass. Social vehicles and buses had to take a bypass from Yanqi Lake North Road, and no motor vehicles were allowed to park along the bypassed route.

From 0:00 on May 15 to the end of the activity, from the east entrance of Yanqi Lake North Road, via Yanqi Lake North Road to Shentangyu Road, except for the vehicles with the special certificates of "Summit Forum", other vehicles were forbidden to pass. Social vehicles and buses had to bypass through Jingjia Road.

(2) **Traffic control measures are taken for roads around the National Convention Center of Chaoyang District.** From 0:00 on May 10 to the end of the activity on May 15, the roads in the area from Beichen East Bridge, via North Fourth Ring Road, Beichen West Road, Datun North Road, and Beichen East Road to Beichen East Bridge (excluding the above roads), except for the vehicles with the special certificates of "Summit Forum", other vehicles were forbidden to pass. Social vehicles and buses could bypass Datun North Road, Datun Road underground tunnel, Huizhong Road underground tunnel, Beichen West Road and Beichen East Road.

From 6:00 on May 14 to the end of the activity, except for vehicles with special certificates of "Summit Forum", other vehicles were forbidden to pass through Beichen East Road, Beichen West Road, Kehui South Road, Datun North Road, Datun Road underground tunnel, Huizhong Road underground tunnel and Minzu Garden Road. Social vehicles and buses must bypass Anli Road, Beijing Tibet expressway, North Fifth Ring Road and North Fourth Ring Road.

Using the big data to restrict traffic is an important experiment in Beijing which had begun since the Olympics. Based on data rather than experience could solve traffic problems more effectively, which is also the main direction of traffic control for major activities in the future. While ensuring the normal operation of urban traffic, it could also show the world leaders a new look of Beijing, the development of Beijing, and even the fact that China is bound to be more mature.

9.4.2 *Application of Beijing intelligent transportation*

At the same time when big data carries out traffic restriction, a large number of new big data transportation emerges. Beijing traffic is optimized from details, and new traffic characteristics of Beijing are displayed during the One-Belt-One-Road International Forum.

(1) **Mobile phone for transportation services.** In 2017, the mobile big data transportation applications in Beijing were launched in an all-round way. Citizens could not only take bus and metro pass through mobile phones but also pay fees intelligently, query road conditions, and bus arrival information through mobile APPs. At the same time, "Beijing traffic APP", "customized bus APP", and other software had opened the channels of road traffic and aviation, and the information about traffic can be queried with one click, greatly improving the convenience of traffic operation.

(2) **Continuous improvement of TOCC system.** In 2011, the Beijing traffic operation monitoring and dispatching center officially opened, referred to as TOCC system. By 2017, the system had been in mature operation and played a positive role in the urban traffic planning. By December 2017, TOCC system had included more than 6,000 static and dynamic data and 60,000 road videos, with data storage reaching 20T. It can be said that the big data and cloud computing provide a strong guarantee. In the future, traffic management will constantly retrieve information from these data, establish a traffic simulation system, start with predictability and initiative, make more judgments on traffic, and send decisions to the relevant departments for fast processing.

(3) **Intelligent prediction system.** The intelligent prediction system is an important exploration of Beijing transportation. The system is widely used in all expressways and main roads in Beijing. It collects the traffic flow of the road in 24 hours, and directly displays the state of the road condition in an intuitive graphic mode through the big data analysis. Once the historical normal value is exceeded, the system will give a warning, which provides an effective reference for the next decision-making of grassroots traffic management departments. At the same time, through the video image recognition technology and automatic detection of traffic accidents, traffic accidents can be handled quickly to ensure the smooth road.

As the capital of China, Beijing is an important exhibition of China's new image, and also carries a series of important international conferences. Therefore, as Beijing's big data transportation continues to improve, it will also radiate to the neighboring provinces and cities so as to improve China's overall traffic outlook. It can be predicted that in the future, Beijing will have more space for big data traffic improvement, so as to ensure the effective operation of this international city.

Chapter 10

Big Data Transportation and Traffic Management

The traffic management is a major issue in transportation and is naturally needed when there is transportation. It involves all aspects of the traffic process, including the management of major traffic facilities and traffic information, as well as the management of technological change, management mode, and organization mode. All round management needs strong real-time response and analysis ability, and the big data is just in line with these requirements.

10.1 State Diagnosis of Big Data Transportation

For the state diagnosis of transportation by the big data, it relies on various information collection, then carries out multidimensional data analysis, finally makes complete data with the help of historical records, current traffic conditions, and so on. Compared with the early stage of the big data development, the content of state diagnosis is too vague and difficult to understand. Now, the current big data diagnosis will form a complete report, and be displayed in the form of charts and other intuitive ways, which can be effectively applied. Therefore, more and more cities integrate the transportation big data effectively form a complete big data diagnosis system and improve the operation ability.

We take Nanjing as an example. As early as 2012, Nanjing had begun the construction of the big data information of transportation. At that time, the core project was "two cards and one center", that is, the citizen card, vehicle card, and government cloud computing center. With the continuous upgrading of big data from 2013, these core projects are integrated into "two networks and one platform", namely, LTE government private network, Internet of Things, and citizen integrated service platform, effectively improving the state diagnosis ability of the big data and effectively communicating various kinds of information.

Especially for the public and service platform, this platform will not only provide relevant data and suggestions to the public transportation department but also is open to the citizens. Through real-time road conditions, citizens can get specific information of each station and road, get the best road suggestions in advance, and ensure the smooth traffic flow of the road. This state diagnosis is the future trend of the traffic management, which can not only make the data more accurate but also effectively reduce the cost of traffic supervision and operation.

In terms of technical means, how does the big data transportation carry out these state diagnosis?

10.1.1 *Operation of information collection module*

In order to carry out state diagnosis, the data collection must be carried out first, which is the basis of big data operation. At present, there are two modes of information collection: static mode and dynamic mode.

In most cities, static information collection has been opened. The characteristic of this method is to capture the corresponding data quickly when the car passes through the video monitoring device or induction loop fixed on the road so as to realize the vehicle information collection. For example, thousands of information collectors installed on Beijing expressway and main road belong to the static collection equipment. This kind of equipment is characterized by its low price and simple fixed installation and maintenance, but its disadvantage lies in its limited flexibility. Once the vehicle evades the monitoring range, it will lead to the failure of data collection. Therefore, the number of static information collection equipment is often large.

Compared with the disadvantages of the static information collection, the dynamic information collection equipment is more flexible. For example, the collection equipment installed on the traffic vehicle and supervision vehicle will collect the data through magnetic frequency, photoelectric, weight sensor, and other methods, which can effectively collect the real-time running speed of the vehicle. The data are in flux, so projections of congestion are more accurate.

The accurate data are the premise of the traffic management and the basis of making accurate judgment for big data. Therefore, a combination mode of "static + dynamic" should be formed to ensure the accuracy of data sampling.

The intelligent bus is a typical combination mode of "static + dynamic" information collection. The new intelligent bus system can continuously collect the driving condition of surrounding vehicles during driving, and quickly judge the condition of the road ahead. Meanwhile, it can also conduct complete static data capture for the bus stops and traffic intersections to provide the data support for the next traffic management. Therefore, to promote the use of intelligent bus is not only for low-carbon life but also for urban road optimization.

10.1.2 *Application of big data analysis module*

When the collection of the big data information is completed, the relevant information will enter the big data platform and the next step of analysis. At present, the third-party big data companies are developing rapidly, so the analysis mode of big data transportation products of different brands is different. But the universal feature is that the information from different channels will be effectively summarized and traffic information data model will be established. Then, these data form image video, and through the deep analysis of the big data platform classification, the valuable information data can be obtained.

The more complete the big data platforms, the higher the accuracy of this analysis will be. Even if some data are fuzzy, the big data platforms will make effective mathematical modeling through historical records, self-contained database, and other ways so as to improve the data analysis.

At present, most of the third-party big data enterprises serving the traffic field have established a complete analysis module, and the Internet giants such as Baidu and Alibaba are also constantly marching into relevant fields, creating a blossoming state of affairs. Local transportation departments should try their best to choose enterprises with rich market experience to cooperate. The first is to ensure the safety of big data platform, and the second is to manage the traffic in the region with the help of a complete database. For example, Alibaba's big data brand has stored 100 PB of data content as of 2014, which is equivalent to 40,000 Seattle Central libraries and 58 billion books. Especially for the traffic development of the new area, with the help of data files of other regions, data supplement can be made for this area, and more reasonable suggestions can be provided for the planning of roads and network lines.

10.1.3 *Application of traffic data processing module*

After the data collection and analysis, the next step will be the final step — processing. Processing is directly related to the state diagnosis of big data transportation so as to make targeted traffic adjustment. Generally speaking, the traffic data processing module will process the distributed data and make the final decision. For example, when there is obvious congestion at an intersection, the data will be analyzed quickly and then a decision will be made: optimize the duration of the signal lights, give priority to over vehicles from the south to the north, and the east–west vehicles can wait appropriately.

The processing of traffic data is not only to adjust the details of traffic management but also to accumulate data and forecast other surrounding areas. Through the analysis of traffic abnormal behavior and the comparison of relevant data in the database, the traffic department can master the road conditions in a period of time, so as to make a more scientific plan. These data seem to be disordered, but through the optimization of big data classification, it can be directly applied to work.

In the final analysis, for the traffic condition diagnosis, the big data need to be carried out through three stages including information collection, information analysis, and information processing so as to create a good traffic network. With the continuous upgrading of the big data technology, more and more traffic related content will be included, such as the flights, high-speed railway, urban construction, and urban maintenance. The richer the data, the more accurate the state diagnosis will be. In the future, the development of big data transportation lies in this. By making full use of the big data technology, efficient and safe traffic management mode will be realized.

10.2 Information Analysis of Big Data Transportation

With the rapid development of urbanization in China, most cities, even rural areas, are experiencing serious traffic jams. The traffic jam not only causes road congestion but also causes serious waste of traffic resources and even affects economic development. In the United States, for example, every family with a car loses $1700 a year due to traffic jams. China is in a period of rapid increase in the number of cars. In addition, the urban and rural reconstruction will also lead to economic waste due to poor traffic.

Therefore, the application of big data transportation focuses on dealing with poor traffic, traffic jam, and road optimization. The reasons of traffic jam in different areas are different, so we must use big data to analyze the information. Only by finding the root cause of the problem can we treat it.

10.2.1 *Deficiency of traditional information analysis*

There are many reasons for road congestion, some of which are due to excessive traffic flow, resulting in exceeding the carrying capacity of the road; due to the imperfect infrastructure, causing the traffic flow to be unable to run normally; due to the lack of surrounding parking lots, causing the car owners to park recklessly; and due to the development of urban construction, leading to more occupying-road construction and resulting in road congestion.

Of course, the traffic jams in many areas are multifaceted. The aforementioned reasons work together. Therefore, a single information analysis is difficult to develop an effective solution. However, the traditional traffic information analysis, due to the lag of technical capacity, leads to obvious loopholes in information analysis.

(1) **The data is not comprehensive enough.** In most cities, the traffic monitoring system often covers the main roads, neglecting the data monitoring of the branch roads and the fork roads. The information received cannot fully reflect the traffic situation, so it is difficult to formulate effective strategies.
(2) **The random sampling is insufficient.** The traditional traffic information analysis usually relies on random sampling. But because it needs a lot of manpower and financial resources, the sample is insufficient, random, and the data is not representative, so the analysis naturally has problems.
(3) **Geographical restrictions.** In the past, there was less information flow between the departments, which brought limitations to the collection of traffic data. For example, the reconstruction of urban

buildings belongs to the responsibility of the ministry of housing and urban rural development, and the transportation department cannot get the relevant data, so it is difficult to conduct traffic diversion, resulting in traffic congestion around. Without the source of information, it is difficult to make correct analysis and prediction.

10.2.2 *Information analysis and promotion of big data model*

When the transportation and big data are combined, the shortcomings of traditional information analysis can be solved. Because, under the big data mode, the information analysis will present the following characteristics:

(1) **All data monitoring.** The big data recognize the intelligent equipment distributed in all corners of the city and collect information uniformly, which pay more attention to all traffic data than random samples.
(2) **Enrichment of data.** In the process of information analysis, the big data are not limited to a single data. At the same time, due to the large amount of storage, data analysis is more abundant. With the combination of static and dynamic data, the information analysis is more accurate and valuable.
(3) **Information analysis in real time.** The characteristic of big data is that the data can enter the database in the first place, and the data can be analyzed dynamically in real time, avoiding the lag of data. In this way, we can know the traffic operation status in a short time so as to make an effective traffic plan.
(4) **Micro data collection.** Although the big data focus on the overall situation, the collection of data is from the micro, so the analysis can be more targeted. For example, for the analysis of bus transportation information, we will start with each passenger's bus IC card, and combine the mobile GPS data. The data is accurate to each person and each time node.

10.2.3 *Analysis and application of big data transportation information in Guiyang*

With the construction of intelligent transportation system in every city becoming more and more complete, nowadays, the analysis of big data transportation information is in-depth. Microwave, coil, GPS, license plate, bus IC card, and mobile APP all form the effective data sources. With the help of these data and information, the traffic department can effectively grasp the traffic information, metro and bus information, and quickly develop adjusted plans.

It is through such big data information analysis that Guiyang city has carried out a drastic transformation of urban transportation. Guiyang is located in the mountainous area, compared with the plain city, and the basic traffic construction is relatively backward. Although in recent years, the construction efforts have been increasing, with the rapid development of population, motor vehicles and real estate, problems such as traffic jams and parking difficulties have become increasingly prominent.

In order to solve the related problems, in 2014, Guiyang transport department introduced big database software, such as MySQL database and began to screen big data for 269 traffic monitoring points. As a result, this big data application has achieved very good results. From September 1–13, 2014, the big data accumulated 154 million pieces of original data. Through the screening of big database, the traffic characteristics of Guiyang were quickly discovered.

For example, according to the big data survey of "the intersection of Huangjin Road and Zaoshan Road", it is found that the traffic volume of this road section increases significantly at 6:00 every morning, reaching the peak at 8:00–9:00, during which nearly 8,000 vehicles pass. From 10:00 to 19:00, the traffic, which has been at its peak, has weakened slightly. At 22:00, the small peak appears again.

In the past, such data analysis could not be achieved simply by relying on the investigation of grass-roots staff. But with the powerful computing power of big data, hundreds of monitoring equipment work together and soon find the traffic data of each intersection in Guiyang. With this data, the transformation of next urban transportation can be accurate and effective.

After several days of the big data analysis, Guiyang transport department found some problems. For example, the ownership of private cars is too high, accounting for more than 70%, which means that there are obvious deficiencies in Guiyang's public transport. In the future, the development of public transport is bound to be focused so as to effectively alleviate the problem of traffic congestion. Especially for the monitoring of the first ring road and the section within the first ring road, it is found that the intersections here are the main places for congestion and the main directions to solve the problems in the future. Meanwhile, compared with

Figure 10.1 Smart transportation system of Guiyang.

Source: Guiyang Network.

other cities, Guiyang is not a simple city having morning rush-hour congestion. The congestion in Guiyang occurs all day and before midnight, so copying other cities is not effective (see Figure 10.1).

With such data analysis, Guiyang's transformation of urban transportation must be targeted, rather than blindly increasing buses or widening roads.

Such a big data analysis mode can be effectively launched in every city, providing the most direct data for the transportation. For example, the bus card can accumulate massive data of passenger travel, from which data analysis will find the main direction and centralized stations of passengers so as to establish a public transport model, effectively formulate the response plans and scientifically allocate the transport capacity.

10.3 Technological Change of Big Data Transportation

The combination of big data and transportation is not immutable. With the continuous improvement of big data technology, the technology of traffic management is also in the process of continuous change so as to meet the development of the times. Therefore, for the grass-roots traffic department, it is necessary to continuously upgrade the big data technology and purchase new big data detection equipment, so that the work can be continuously improved and always conform to the development of the times.

10.3.1 *Upgrade of hardware equipment*

The early detection of big data is often just an upgrade of the traditional traffic equipment with video shooting function and data return function. But with the continuous innovation of technology, the hardware equipment has been significantly improved. In particular, the use of these hardware equipment provides more abundant means for big data detection.

(1) **Internet of things sensor.** Compared with traditional sensors, the information capture content of Internet of Things sensor is greatly improved. Traffic management department embeds Internet of Things sensor on the road, which can observe the real-time vehicle traffic and passenger flow information on the mobile, and supervise the vehicle exhaust emissions. The Internet of Things sensor can achieve synchronous data collection for multiple vehicles, which is more efficient. It lays the foundation for commanding the traffic management system and collects the data more efficiently.
(2) **Application of satellite map.** In the past, the satellite map data were not released to the public, and most of them belonged to military applications. But with the wide application of the big data in commercial and civil fields, satellite map can also be effectively used in the traffic management. Therefore, the traffic management department should access the satellite map as soon as possible, and formulate corresponding strategies through the real-time capture of traffic conditions by the satellite so as to carry out traffic planning and road planning.
(3) **Application of smartphone.** For the public, vigorously advocating the big data application of smartphones can also effectively solve the traffic problems. Smartphones have been widely used. They can easily link and interact with related platforms of traffic management departments, such as WeChat, Weibo, and related APPs. At the same time, the application of GPS positioning and BeiDou positioning system enables map suppliers to play a good role in their own big data

application, pushing the road congestion and travel trend for passengers and car owners in the first place. These information can optimize the road.

Based on these three technological changes, the intelligent traffic lights, big data ticketing platform, and other technologies have been fully mature. Both the enterprises and government departments have begun to use these technology reform dividends to transform the transportation industry. Actively introducing advanced technology into the traffic system will bring a new situation to traffic management.

10.3.2 *New big data technology change*

The big data technology is still being updated, bringing new changes to traffic management. For example, the current mainstream intelligent big data transportation technology, mainly uses the intersection monitoring equipment and electronic police. It still belongs to the traditional technology to achieve the capture of vehicles, the recognition of license plate number, and other works. Although the vehicle can be analyzed quickly, it is difficult to dig deeper. The traditional algorithm makes the big data transportation meet the bottleneck.

The new big data technology has shown such characteristics: through the photos taken, it can detect the number of annual inspection, the shape of the annual inspection mark, whether there are dangerous things hanging on the car windows, and whether there are illegal acts by drivers. For example, the intelligent traffic recognition algorithm developed by Beijing Tucun Science and Technology Institute can achieve these goals and help the traffic police departments to quickly find the road information. Once the problem is found, the data will be stored as soon as possible, providing strong evidence for the future case handling by public security, and letting drivers pay attention to their own behavior, so as to further ensure the traffic safety.

At the same time, the continuous promotion of "driving perception technology" has also significantly improved the access technology of traffic data. In the future, the relevant equipment will be installed when the car leaves the factory. Through the onboard "Internet", the relevant data of the car will be released to the relevant big data platform in the first instance. Even the driver's emergency brake will be recorded, the surrounding conditions will be photographed quickly and the road hazards will be found at the earliest. The complex driving environment is effectively recorded by perception technology and the report is obtained through the analysis and screening of big data platform. If the driver's emergency brake often occurs in a road section, it means there is a traffic hazard, which will give the data support to the traffic department and urban construction department for later road adjustment.

"Driving perception technology" has not become the mainstream in China at present, but with the continuous efforts of automobile manufacturers, it is believed that this technology will soon become the mainstream configuration of automobiles. Through the development of big data technology, we can see that in the future, the traffic management is not limited to traffic management departments, but all car owners are part of it. Only when their own information is effectively captured can they attract the attention of relevant departments, and finally promote relevant departments to make corresponding adjustments and changes.

The significance of big data lies in this: it is not a simple tool, but a hub connecting traffic management departments, car owners, citizens, cities, and transportation. Through this hub, the information of all parties is effectively summarized, from which the rules are found for accurate solution. Therefore, when everyone can send their own information through big data, the effect of traffic management can be significantly improved. This also means that in the future, the big data upgrade will inevitably be more closely connected with everyone, and be able to effectively capture every detail of the traffic process, even including how long the pedestrian waits for the traffic light, the distance from the bus stop to the front door, and so on. These seemingly useless data can be effectively collected and screened to create a real intelligent and three-dimensional transportation system.

10.3.3 *Shenzhen: Forerunner and leader of the use of big data transportation*

Shenzhen, the youngest, most dynamic and innovative city in China, is also the first city in China to try big data intelligence. Relying on a large number of emerging IT enterprises, Shenzhen's transformation in big data transportation technology has been leading the country. Compared with Beijing, Shanghai, and Guangzhou, the traffic in Shenzhen is also the best. Therefore, it is worthwhile for other cities across the country to learn from Shenzhen's model.

As early as 2000, Shenzhen had established an intelligent traffic command center. Relying on intelligent signal lights, closed-circuit television, intelligent traffic violation management system, Shenzhen's traffic management integrates information, monitoring and command, which can effectively manage the urban transportation.

Subsequently, the big data technology is continuously applied in the field of transportation in Shenzhen. In August 2007, eight enterprises in Shenzhen established the Intellectual Traffic Safety Association of Shenzhen, which built a more efficient and intelligent communication and development platform. The "intelligent traffic 1 + 6" project invested by Shenzhen with 1 billion yuan has made the big data technology get a new upgrade.

The "1" in "intelligent traffic 1 + 6" project is the "one resource sharing platform", which includes the transportation information, traffic police, urban construction and other departments, and all the information has been effectively exchanged. The "6" refers to the six service systems under the platform: traffic monitoring information, traffic management information, road traffic control information, public travel information, traffic command emergency information, and traffic management, and decision-making information. It is by virtue of the "1 + 6" system that Shenzhen's big data transportation has realized a comprehensive upgrade and effectively serves urban transportation.

In order to make better use of the big data, Shenzhen has continuously accelerated the construction of the intelligent comprehensive traffic operation and command center and carried out the construction of subcenters according to the actual situation, aggregating more mass data of public transportation. For the vast majority of cities in China, the Shenzhen model is the best reference, constantly upgrading for the big data transportation technology so as to cope with the growing number of cars and urban population. Science and technology serve people. Every progress of science and technology may bring great improvement to the previous work.

10.4 Organization Model of Big Data Transportation

The application of big data transportation can not only rely on the independent action of the transport department but also must be carried out in combination with multiple departments and actively introduce the advanced technology of the third-party big data enterprises so as to form a perfect organization model. The organization model determines the system and framework of big data. The more mature the organization, the more sufficient the data can be. Otherwise, once the "information island" is formed, the data will inevitably be distorted and cannot effectively serve the traffic management.

10.4.1 *Big data model changes "information island" effect*

The "information island" effect means that the source of information is very narrow, often only limited to individuals and small groups, which has no reference value and is difficult to be applied in a wider range. The traditional traffic organization model only relies on the human resources of the traffic department, the data statistics is very limited, with a strong personal subjective characteristics. It is often the person in charge of the research that determines the final data, so in the past the traffic planning of many cities was not ideal. Today, with the proliferation of vehicles, traffic jams and accidents are frequent.

The organization model of big data is to break the "information island" effect. Taking the automobile sensor as an example, when the vehicle passes a certain intersection, the monitoring equipment will quickly capture the data, and the relevant data will be quickly collected. In this process, not only the traffic department needs to install the corresponding equipment but also the vehicle monitoring agency needs to cooperate and supervise the vehicle to install. At the same time, the urban construction department will reserve the relevant interface for the traffic department in the process of relevant facilities construction, so that the big data can really play an effective role. Therefore, a perfect public transport cloud can be established and all organizations can access it effectively, so that data can be circulated and statistics of data can be more effective so as to meet the needs of traffic management.

10.4.2 *Three-level management organization structure of big data*

The organization model of big data lies not only in the information transmission between different departments but also in the different organization structure of different departments, forming a "three-layer structure". Among them, the communication service provider is the support to complete the final data collection.

(1) **Information layer.** The first layer of three-layer structure is the information layer, whose function is information collection. This requires the cooperation between the traffic department, the satellite data department, and the communication service provider to quickly collect data through the satellite and mobile phone terminals, and then transmit it to the transportation server period through the network.

    Subsequently, the relevant technical personnel conduct data investigation on the specific location and route of vehicles or individuals through various allele algorithms, and store and manage in the cloud service platform, while residents can view relevant information through mobile terminal APPs such as mobile phones and enjoy the convenience brought by big data transportation.

(2) **Network communication layer.** The second layer is the network communication layer. The work of this organization system is to organize the data effectively and transmit the data to the cloud service platform through WLAN, 4G, and other ways. For this, the third-party big data company must play an active role in ensuring the effective operation of the server and avoid information omission, loss, and other problems.

(3) **Cloud service layer.** The third layer is the cloud service layer. The main work of this layer is to code and transform the data, mine the information, and predict the traffic jams. This requires professionals to make use of big data platform to ensure the accuracy and effectiveness of information screening. Especially for the densely populated road sections such as the city center and the business district, the work of this layer is busier, providing perfect path optimization, real-time public transport query and other services, so it needs the cooperation of multiple departments to effectively solve the problem.

10.4.3 *Organization system of big data transportation*

With the continuous upgrading of the big data technology, the big data transportation has formed a complete industrial chain, including data collection, integration, analysis, mining, display and service. In the future, most cities will establish a detailed organization structure according to these five aspects, so as to do a better job in big data transportation. These organizations are not only composed of traffic management department, but also the third-party big data enterprises.

(1) **Data collection.** Most applications of big data transportation in cities often rely on induction coils and microwave radars which constitute the main body of urban transportation monitoring mainly installed and monitored by the traffic management departments. However, these collection equipment are difficult to cover the road network of the entire city, so it needs the intervention of a third-party detection company. Most of these companies have their own set of perfect equipment, will carry out specialized training for staff, and become more skilled in the application of new technology. Therefore, the data collection can be partially outsourced, forming a mode of joint operation of "traffic management department + third-party big data enterprise" so as to improve the universality and professionalism of information collection.

(2) **Data acquisition.** In the past, the traffic data of domestic cities were often mastered by many departments, such as the municipal administration, traffic, urban construction, planning, and traffic police. The application of big data transportation will effectively unify these information, gather, and integrate them in the traffic management department so as to realize the standardization of data collection and storage. The traffic information centers established in different regions are typical "centralized big data" organizations. Only by unifying data management can we find and solve problems from different perspectives. At present, most of the information gathered by the traffic management department is the public transport information, such as buses and metros. In the future, the taxi information, shared car, and shared bicycle information will also be effectively involved and the data will be more informative. Therefore, it is the task of the development trend of big data transportation to strengthen cooperation with these commercial companies and make data information unified as soon as possible.

(3) **Data analysis.** For the analysis of big data, most of them will adopt the mode of cooperation with the third party. Compared with the traffic sector, the third-party companies often have a strong advantage in data analysis. No matter the platform equipment or talent reserve, they far exceed the government departments. For example, Shanghai

Meihui Company, and Shanghai Institute of Electrical Science are excellent data analysis companies in China. The traffic department can inform these companies of its own needs and they will immediately process relevant big data, such as travel information, traffic decision-making information, etc. For example, the data analysis for taxi GPS is the specialty of a third-party company.

(4) **Data thinking.** Pure binding analysis is only the primary application of big data. It is the development goal of big data transportation that makes the data generate analysis and have a better traffic management form. Therefore, the traffic department must carry out in-depth cooperation with the third-party companies and high-efficiency companies, and let big data play a greater role through patents and papers. For example, Tongji University participated in Shenzhen urban transportation simulation system, which greatly improved the planning level of big data platform and made great contribution to the improvement of urban transportation.

(5) **Information service.** The information service will directly face the end users and push the information processed by big data directly to the residents. In this process, the traffic department should cooperate with the third-party information service department to make the information transmission more extensive and accurate. For example, the traffic department publishes complete traffic information through the APP, the campus APP publishes specific measures around the campus, and the car owner APP pushes specific bypass plans so as to form a complete information release system.

It can be seen that the organization model of big data transportation is no longer limited to the traffic management department, but the situation becomes "multi-point cooperation and all-round development". Only by forming such an organization model can big data really complete the practice from theory to practice from information collection, information analysis to information release.

10.5 Management Model of Big Data Transportation

There are many kinds of big data transportation management models, but their essence is the same: that is, through continuous capture of data and integration of a variety of data, the valuable data applications are ultimately found. Different treatment methods are only different in practical means. Especially for the big data platform created by the third-party enterprises, their essence is similar. Therefore, in the application of the big data, as long as we find common ground for management model, we can quickly open up the situation and accurately serve the traffic field.

10.5.1 *Common big data management model*

Generally speaking, the following types of big data management models are the most common management models at present:

(1) **Spatial state analysis.** With the help of all kinds of big data equipment installed on the street, the big data analyze the spatial state, find out the characteristics and changing trend of the traffic flow distribution and monitor the road state so as to make the congestion index, and manage the traffic by adjusting the traffic signal lights. This model is a conventional management model which is applied to daily traffic management.
(2) **Risk management model.** For the high-risk traffic sections, the risk and vulnerability analysis is carried out with the help of big data. Through the comparison of historical data, the problems that are about to occur are predicted, and the risk management mechanism is established. This model is usually used before the temporary large-scale activities.
(3) **Integration model.** The feature of the integration model is to make comprehensive use of positioning, communication, and network technology, especially to predict the risk of major events that may occur, improve the ability of prevention, and ensure the orderly development of traffic. Generally speaking, before the major holidays, the integration model will be opened to cope with the coming traffic.
(4) **Trajectory tracking model.** Trajectory tracking model realizes the search of the same number and firmly locks the trajectory of people and vehicles with the help of the correlation of various information, such as license plate tracking and new license plate tracking. Generally speaking, if there is a major traffic accident and the perpetrators need to be arrested quickly, this big data management model will start.

(5) **System comprehensive analysis model.** Through the modes such as integration of big data and association history, the system comprehensive analysis will carry out perfect traffic processing. When setting the traffic conditions in the new area, the system comprehensive analysis model will play a positive role in road planning in advance to avoid risks.

These common management models cover all aspects of traffic and effectively improve the application of big data. The accurate management system is built by category, which will greatly guarantee the traffic operation.

10.5.2 *Upgrade of big data transportation management model*

Five basic management models form the foundation of big data transportation management. On this basis, the big data transportation management is upgraded so that the big data can play a more effective role. In view of the current traffic situation in China, the upgrading direction of big data transportation management should focus on the following points:

(1) **Cross-region management system.** The traffic departments in each region have certain administrative autonomy rights, so the information flow among regions is not smooth. The problem is particularly acute among provinces and cities. Therefore, big data transportation should break through such information barriers and realize resource sharing so as to optimize the traffic situation from a macro perspective.
(2) **Giving full play to the advantages of aggregation.** Because the traffic departments at all levels implement the model of administrative regional autonomy, and different transportation themes are in the charge of different departments, the data are often too scattered and the efficiency cannot be improved. The introduction of big data management model is to break through the information barriers between various departments, such as the data of bus companies, metro companies, taxi companies and shared bicycle companies, so that each station can learn from each other when it is established so as to improve the efficiency. This is not only the optimization of traffic

management but also the optimization of cooperation model between different departments, so that the administrative efficiency can be effectively improved.

(3) **Optimization of resource allocation.** The lack of resource allocation is the short board of traditional traffic management system. Some departments are so well staffed and equipped of monitoring, even there is a lack of application of resources. Some departments are unable to effectively carry out traffic management work and make full use of resources due to the lack of human and financial resources. With the help of big data aggregation, resources can be effectively centralized to make up for the uneven distribution.

It can be seen that the upgrade of big data management model lies in "resource sharing and department cooperation". Only by establishing a perfect management system and carrying out optimization of comprehensive organization can big data play a further role and improve the efficiency of traffic management.

10.5.3 *Hangzhou model: Comprehensive utilization of big data*

At present, the big data transportation management model of Hangzhou in Zhejiang Province has gradually become clear, forming a new management model of "one center, three systems", and effectively integrating the traffic departments in the city. "One center, three systems" refers to the traffic command center, traffic management information system, traffic control system and traffic engineering information system.

This management model enables the traffic management departments of each urban area to cooperate effectively and the traffic police detachment becomes the main force of ground traffic service. According to the information of the platform, it will implement the hierarchical early warning and intervention mechanism at the main urban roads and intersections. At the same time, the digital service room of the traffic police squadron carries out network patrol and duty. Through the application of big data, the traffic problems are found at the computer end, and the coverage of road management is effectively expanded. When the problem occurs in a concentrated way, the joint action shall be carried out immediately, and the traffic police detachment, urban management, urban construction, and other departments shall respond quickly to solve the road traffic problem, which will greatly alleviate the traffic congestion and ensure the safety and smoothness of the traffic flow on the road.

Hangzhou big data transportation management model also actively introduces the power of the third-party enterprises. The data of Didi, Baidu Map, and others are also imported into Hangzhou's big data transportation platform, which is released by Hangzhou traffic police department. For example, with the help of big data statistics of third-party enterprises, Hangzhou traffic management department found that by the end of February 2017, the number of female drivers in Hangzhou had been 1.3964 million, accounting for 36.3% of the total number of motor vehicle drivers, which could provide data support for the formulation of traffic laws and regulations. The establishment of parking lots, drainage, and planning routes can take care of female groups, which is more in line with the characteristics of traffic in Hangzhou.

In December 2017, another big data transportation event happened in Hangzhou. On December 2, 23 big data enterprises set up the Zhejiang Transportation Big Data Maker Space, featuring "Internet + transportation", to serve the transportation of Hangzhou and even the whole Zhejiang province. Among them, there are not only the Zhejiang Transportation Information Center, the Settlement Center of Zhejiang Highway Bureau, the Zhejiang Tourism Information Center, Zhejiang High Speed Information Engineering Technology Co., Ltd. and other institutions but also Baidu, Huawei, Alibaba Cloud, ProLogis, Beijing Yihualu, Cybernaut, and three major operators to build a complete new model of big data management. In the future, the big data management model will inevitably present the mentality of "departments and institutions + third-party enterprises" to work together and effectively combine various forces to find the key to the development of urban transportation.

10.6 Credit Model of Big Data Transportation

The operation of transportation big data also needs to be supported by credit. This is because, regardless of the data collection, screening, analysis, and release, the grassroots people are often in the receiving end. If the information frequently deviates, it will inevitably lead to people's distrust on big data transportation and affect the operation of big data.

So, from what angle should we solve the trust problem of big data transportation?

10.6.1 *Multi-channels and professional releasing*

When releasing big data, the traffic department should cooperate with other channels besides its own platform, especially the official platform and certification platform such as the local traffic radio station, television station, onboard terminal, vehicle rolling display screen, variable information board, and traffic warning signs. Big data should be connected with these channels in time to release the information in the first time so as to let the public know the relevant trends in the first time.

In addition, in public places, the traffic department should also establish cooperative relationships with regional WeChat official platform and Weibo provide relevant information to operators and release data through their platforms. We must choose the certified platform to cooperate so as to ensure that the information is truly and effectively pushed to the public, and enhance the credibility of data release. Once there is lack of cooperation with these channels, it will inevitably lead to the rapid spread of false information and rumor information, and the value of big data will not be reflected, which will further affect the credit value of the traffic management department.

10.6.2 *Easy to understand and professional interpretation*

For big data transportation information, because it involves professionalism, many contents must be reprocessed so that they can be easily understood and comprehended by the public. However, easy to understand does not mean "unprofessional". Otherwise, it is easy to cause public misunderstanding and mistrust of the big data information.

Therefore, in the process of easy understanding of the big data, the traffic department should also interpret it from a professional perspective, so that the public can understand the operation mechanism of big data and be convinced. For example, when the data are released, the traffic flow analysis, time-out prediction, and time identification should be emphasized, and the public can see the problem through historical record comparison. At the same time, the traffic conditions should be analyzed in combination with static photos and dynamic videos so that users can feel the valuable information, and also see how big data mining information, how to analyze information, and how to make decisions, so that the credit value of big data will greatly improve.

Once there is trust in the big data, the public will actively participate in the process of information collection. For example, most mobile APPs have positioning functions. When the public realizes that positioning can effectively carry out data transmission and further optimize traffic routes, they will also actively upload geographic location and enable monitoring functions so as to promote the data collection of big data more perfect.

10.6.3 *Uniting the public security department to crack down on rumors*

The traffic is related to the basic life of the public, which is the focus of public attention, so it is also the focus of rumors. If the traffic management department does not deal with these rumors, it will easily lead to the spread of false information, greatly influence the normal traffic information release, and doubt the big data traffic management. Therefore, the traffic department should take the initiative to cooperate with the public security department to crack down on traffic rumors and avoid the further spread of false information.

For example, in July 2014, Guangning County, Zhaoqing City, Guangdong Province suddenly had news that four people were killed in a major traffic accident the day before in the path section of Nanjie Town, Guangning County, with three pictures. The news spread rapidly in the local area, and a large number of netizens reprinted it. The local people questioned the work of the traffic police. Subsequently, the local traffic police department quickly intervened and found that this was a network rumor. In order to make a grandstanding, the suspect released false information in his QQ space. The traffic police department and the public security department quickly linked up to detain the suspect according to law, finally dispelled the rumors, and the image of the traffic department was restored.

For this kind of incident, the traffic department must treat the data strictly, otherwise it will easily make the public have a negative impression on the traffic department. Once this happens, even when accurate big data information is released, it would be difficult to obtain the trust from the public and the significance of big data transportation management would no longer exist.

Chapter 11

Future Development of Big Data Transportation

The big data transportation has developed for many years and gradually matured, but there are also many challenges. Moreover, with the industry segmentation, in order to serve the public well in the future, the transportation big data need continuous improvement of technology and continuous integration of multiple levels of content and technology, so that in the future, the big data transportation will be more intelligent and humanized.

11.1 Personalized Service of Big Data Transportation Industry

Although the China's big data transportation industry started off quite late because of basic livelihood issues involved, and the rapid layout of major commercial companies, the industry shows a rapid development momentum. In the past, China's big data transportation needed to rely on the technology of the United States and other developed countries, but now China has become the leader of the big data transportation industry, with various new technologies and concepts emerging frequently. Looking at the development of the industry, the big data transportation industry in the future will further penetrate into the personalized service, and even form a "one person's transportation big data" mode to achieve "customized exclusive service".

Some of the following personalized services have begun to enter the R&D stage, while others are still at the project approval stage. They outline the development direction of big data transportation industry in the future.

11.1.1 *Intelligent vehicle–road collaboration technology*

At present, China's automobile market has started the attempt of "intelligent system". A series of independent brands, such as BYD and Chang'an, have carried out the attempt of vehicle-road collaboration technology on the new vehicle models. For example, the "zebra technology" introduced by MG Automobile integrates a variety of navigation modes, which can display the current road congestion in real time, effectively assist the driver to re-plan the road and optimize the road resources. This is a typical "vehicle–road collaboration" attempt.

However, although some manufacturers have begun to try, the large-scale applications have not been fully launched. Especially in the process of high-speed driving, the perception of long-distance environment is not yet fully mature. At the same time, the standard and specification system of vehicle–road collaboration has not been established, which leads to the confusion of technology and bad experience for users when the system is launched.

It can be predicted that in the next few years, the "intelligent vehicle–road collaboration technology" will enter the blowout period, and even become one of the standard automobile accessories. This technology can effectively combine with drivers, capture drivers' personal driving habits, and combine with big data to customize and optimize driving mode and route planning.

11.1.2 *Autopilot technology*

The autopilot technology is the most concerned new transportation change in recent years. At present, most of the world's automobile enterprises have begun to carry out the research and development of autopilot technology, and China's Baidu has successfully launched its own autopilot vehicles, which means that the autopilot has taken a key step.

Compared with the traditional driving mode, the autopilot technology completely relies on big data for operation. It is a high-end artificial intelligence, which can judge the road conditions in real time based on the massive map data. Compared with manual driving, the autopilot technology will play an extreme role in route, driving mode and emergency avoidance, making big data produce the greatest effect. Therefore in the future, the autopilot technology will gradually replace the traditional driving mode, and the big data will become the first "natural person" of autopilot.

11.1.3 *Internet vehicle*

Compared with the vehicle–road collaboration technology and the automatic driving technology, the Internet vehicles have been widely used. The characteristic of Internet vehicle is to implant Internet skills into traditional driving mode, such as voice real-time broadcast of road information, the mobile phone and the vehicle association in the crisis road driving vehicle through voice, fast payment when driving into the highway, and intelligent payment in the parking lot. Many brands of vehicle implanted the mobile payment, which is an effective attempt of Internet technology.

For example, the physical examination for vehicles is an effective application of big data through the Internet. Through the big data, the Internet vehicles collect the data of all users' car use, analyze the vehicle's excellence, maintenance cycle and tire pressure and send the information to the owners through onboard screen or mobile app, etc. In this way, the owners will fully understand the condition of their vehicles and conduct timely maintenance and repair to avoid sudden damage to the vehicle on the road. When the vehicle is in a healthy state, the road accidents will also greatly decrease.

11.1.4 *Rapid promotion of ETC technology*

At present, the ETC technology is actively and effectively promoted in many provinces. It is also the embodiment of personalized services in big data transportation industry. The ETC, the electronic toll collection system, is currently the most advanced toll collection method in the world. When the vehicle is installed with an ETC card, it can use the microwave special short-range communication between the microwave antennas on the ETC lane in front of the high-speed toll station, and use the computer networking technology to carry out background settlement processing with the bank. The vehicle can complete the payment without waiting, which greatly alleviates the problem of queuing payment at the high-speed intersection.

There is no doubt that the ETC technology can be used in traffic optimization, not only in high-speed intersections but also in scenarios suitable for payment. In the future, the ETC will further play the role of "traffic resource optimization", deeply integrate with public transportation card and electronic transportation administration card and realize the purpose of "three cards in one". In this way, the mode of "transportation + big data" will be further formed. The card can meet many purposes, such as the driving, riding, and paying fees, greatly improving the traffic efficiency and alleviating the problems of traffic congestion and resource waste.

11.1.5 *Further integration with individuals*

The so-called personalized service must be combined with individuals to highlight "non-popularity" so as to meet the definition of personality. For personalized exploration of big data transportation industry, it must be precise to individuals. With the accumulation of the big data, we can analyze the daily travel habits, shopping habits, and living habits of travelers so as to make accurate suggestions and judgments. Even the big data can provide a unique way to travel by analyzing travelers' personality, physical condition, and environment. For example, for an old man, the big data will quickly push the travel plan of "couchette" according to his physical condition and living habits; for young people who are keen to explore, big data will push the combination of "plane + car rental". This personalized service can not only provide travelers with unique travel experience but also make efficient use of travel time, optimize various modes of transportation and realize the effective use of traffic resources.

11.1.6 *"Going to the rural areas" for Internet taxi*

Compared with urban areas, the transportation problems in rural areas are smaller, but with the continuous development of urbanization, the transportation problems in rural areas are increasingly prominent. In urban areas, the Internet taxi has become the mainstream. It can effectively optimize the transportation resources and let the empty driving vehicles quickly find the people who need to use the car through the big data information transmission. The optimization of transportation is very obvious. In the future, this taxi mode will gradually go to the rural areas.

For example, Guizhou Province has already begun to try this. Guizhou launched the "Tongcuncun APP" for rural areas, through which villagers can call surrounding taxis and buses. At the same time, the APP will also show the departure time, stops, seats, and other information in detail, so that villagers can directly purchase tickets through the system. This rural version of the Internet car hailing mode of "Tongcuncun" has been widely concerned by the capital market and the big data transportation industry has begun to penetrate in rural areas.

In addition to the aforementioned contents, the big data transportation industry will further solve the problem of travel traffic from the content, such as precise recommendation push for personal vehicle purchase. When the big data transportation industry is becoming more and more personalized, the resources can be effectively complementary and the road problems can be effectively optimized, thus creating a new transportation operation system.

11.2 Expansion Trend of Big Data Transportation Industry

Today, the big data transportation has formed a complete industrial system, which has formed a large-scale feature in theoretical research, product production, platform service, etc. With the further growth of the number of cars and the rapid development of urbanization, the big data transportation industry will further expand and gradually penetrate into every detail of life. In the future, what kind of new industrial upgrading mode will big data transportation have?

11.2.1 *The general trend of cross-border*

The resident travel relies heavily on taxis. However, most cities have a limited number of taxis and the supply and demand cannot be balanced. Especially in the peak period, the taxis are hard to find, and even some areas have seen the situation of private price hike of taxis in the peak period. This phenomenon not only leads to poor user experience but also greatly interferes with the normal transportation development.

Through the analysis of the big data, we can see a phenomenon that in the field of transportation, the car is the entrance of service. If we do not solve the problem of cars, we cannot solve the problem of transportation.

Therefore, in the future, the transportation industry will take "car" as a breakthrough point and form a cross-border industry development mode through the link of big data. The car is not only a simple means of travel but also the entrance of life service. For example, cars can collect data from catering, housing, travel, shopping, and entertainment to understand the life characteristics of the users. When the passengers step into the taxi, the big data system will quickly carry out data analysis, and push the information of passengers through the car so as to meet the "catering, clothing, housing and travel" service. All scenes are associated, while the cars are the exit of association, and big data will integrate everything.

It can be predicted that the big data transportation industry will continue to penetrate into entertainment, catering and life services, complete cross-border integration and make the definition of car generate a new change.

11.2.2 *The further development of sharing culture*

Since 2016, the shared culture has attracted more and more attention, and the most attention is "shared transportation" — from the initial shared bicycle to today's shared car, the focus of capital and people's attention are all projected on the transportation. Regardless of shared bicycles or cars, they have only one purpose to meet, that is, the convenience of residents' travel and optimization of social resources. A vehicle can be used by many people in turn so as to maximize the use of transportation resource which is a positive exploration of low-carbon life and urban transportation improvement.

The deepening of the shared transportation culture indicates that there are still greater opportunities for the industry to be explored. The National Information Center has also made a prediction on shared cars that every shared car can reduce the purchase behavior of 13 cars. The statistics show that at present, a car is used for about 1.5 hours a day and parks in the parking space for 20–23 hours. The time-sharing lease can make a car run for 6–9 hours on the road, greatly improving the parking efficiency. This is the most positive improvement for urban transportation. When the shared culture is accepted by more people, the efficiency of transportation will be improved significantly, and the problem of illegal road occupation will be solved greatly. Therefore, the transportation authorities in Beijing, Shanghai, Guangzhou and Shenzhen, as well as other second-tier cities, have shown an encouraging attitude toward this new mode of travel.

Whether shared bicycles or cars, they are representatives of green travel. The mainstream model of shared car is the new energy electric vehicle, which can effectively reduce the air pollution and the number of motor vehicles. It is an important supplement to the traditional transportation. When solving the problem of travel, it can reduce the congested road traffic. When the majority of people choose to share cars and give up buying cars, the urban transportation will improve obviously.

At the same time, compared with private cars, shared bicycles and cars are more active in the application of the big data. Only by introducing the big data technology, can users quickly find vehicles to use. In the process of application, the big data are still running, which will record the user's driving trajectory, driving frequency and driving time, and further enrich the big database. These data can be effectively used by the transportation authorities to bring first-hand reference data to the transportation management.

Therefore, although the shared economy is now "in full swing" and various enterprises are competing for the market, its market potential determines that there are more opportunities to be found here. This is one of the key directions for the expansion of big data transportation industry.

11.2.3 *New opportunities of big data intelligent navigation industry*

The navigation is also an important part of big data transportation industry. There is no doubt about the role of navigation, which can provide the most direct traffic information services for car owners, optimize road driving, effectively reduce the probability of traffic accidents and relieve traffic pressure.

The new intelligent big data navigation system can not only provide map services but also effectively crack down all kinds of traffic violations. For example, it can inform the car owner in real time whether the car is overspeeding; whether parking is available in the current area; what kind of traffic accident exists ahead, and how long the waiting time is expected. At present, many vehicles have been installed with this intelligent navigation equipment, which has played a significant role in road traffic optimization.

In the future, the function of big data intelligent navigation will be further developed to realize personalized service.

(1) **Connecting with big data of transportation authorities.** The navigation system will be involved in the big data platforms of the transportation authorities. Once it is found that it is about to overspeed, it will immediately give an early warning prompt. At the same time, the transportation authorities can directly push the information to the intelligent navigation terminal to give effective suggestions to the owners.
(2) **More advanced speech recognition technology.** Most of the current navigation systems are poor for speech recognition, which leads the car owners to stop and input, which greatly affects the smooth traffic flow on road. In the future, the intelligent navigation system will greatly improve the ability of speech recognition, even effectively recognize local accents, so that car owners can quickly input speech so as to shorten the road parking time and avoid the traffic congestion caused by stopping.

While the intelligent navigation industry has been effectively broadened, China also further supports the big data intelligent navigation industry from the policy level. In 2013, the General Office of the State Council issued the Medium and Long-Term Development Plan of the National Satellite Navigation Industry, which proposed "by 2020, the scale of China's satellite navigation industry will exceed 400 billion yuan, BeiDou and its compatible products will be widely used in important industries and key fields of the national economy and gradually popularized in the mass consumer market, contributing 60% to the domestic satellite navigation application market, more than 80% to the important application fields and having strong international competitiveness in the global market". This means that with the continuous upgrading of the satellite technology, the navigation, as a derivative industry, will inevitably enter a more vigorous development stage. It is predicted that by 2020, the market scale of China's satellite navigation products and services is expected to reach 400 billion yuan, which is the expansion trend of big data transportation industry.

11.2.4 *Application of big data tachograph*

Tachograph is also one of the industry fields that needs to be explosively developed in the future. The traditional tachograph can only achieve the video shooting function, but the big data tachograph will be directly connected to the cloud database, when the data value of real-time shooting will be effectively used.

For example, when tens of thousands or even hundreds of thousands of cars in a city are equipped with such recorders, the captured information will be immediately received by the large database, which will quickly analyze the operation of urban roads according to the real-time picture of the large database. These data will form effective information, which will be sent to all other car owners at the same time and even automatically set up the navigation destination and driving route to predict the road condition information in the next time period. These information are completely captured and analyzed in real time, so it has great practical value. Therefore, the big data tachograph will not only bring effective help to road traffic but also form a huge commercial income.

In addition, the smart tires generated around the big data belong to a part of big data transportation industry. These industries are now showing a very expanding trend, which will bring new changes to the development of transportation.

11.3 Investment and Financing of Big Data Transportation Industry

The continuous development of the industry is bound to form large-scale investment and financing. The same is true for the big data transportation industry, which is at the stage of tuyere and will inevitably lead to another round of investment and financing. In recent years, a series of news about big data transportation industry all reflect this point.

11.3.1 *Strategic cooperation between China Transinfo and Inrix*

In February 2015, there was an explosion news in the field of big data that *China TransInfo Technology Corporation* signed a strategic cooperation agreement with Inrix, marking a new stage of China's big data transportation development.

Why does China Transinfo choose Inrix for strategic cooperation?

This is determined by the technical capabilities of Inrix. Inrix was founded in 2004 by Brian and Chapman, former employees of Microsoft. Inrix is a transportation intelligent platform dedicated to bringing intelligent data and advanced analysis methods to global transportation problems. By December 2012, it had provided services to enterprises in 32 countries. The partners of Inrix including Audi, German Automobile, Dutch Automobile Association, BMW, Microsoft, Ford, and Toyota are the world's top big data transportation operation brands.

As a result of the establishment of cooperation mode with a series of top automobile brands in the world, Inrix has 100 million vehicles and equipment to collect resources of big data, including taxis, transport vehicles, trucks, etc. With the help of these data, Inrix has established a complete big data information base, which can not only provide traffic route guidance and road condition reminders but also predict global road conditions. The dynamic data help Inrix provide a series of services for customers, even including details such as how to save fuel time for drivers.

Because of this strength, China Transinfo chose to cooperate with Inrix strategically. With the help of the large database capability of Inrix, China Transinfo can quickly explore the big data in the domestic market and accurately serve the transportation; and with the help of China Transinfo, Inrix can also quickly enter the Chinese market and enjoy the dividends brought by the rapid growth of China's transportation.

After the strategic cooperation with Inrix, China Transinfo has carried out rapid business exploration, traffic planning and design in Luoyang, Henan Province, and business expansion to more than 10 other cities. In addition to the traditional road design, flights, water transport and intercity, the roads are all included, realizing the accumulation of resources in the fields of transportation big data realization path, parking search, and service operation.

This event has started the investment and financing boom of the big data transportation industry. More and more enterprises see that big data transportation industry has great potential and rich realization channels like urban road design and planning, electronic station board with PPP mode, O2O realization, value-added services of telecom operators, etc. These services can achieve hundreds of millions of profits so they are favored by capital institutions.

11.3.2 *The financing probability of start-ups increases greatly*

As the development potential of big data transportation industry is unlimited, the financing opportunities for those start-ups with strong technological strength are also greatly increased. For example, in July 2017, Hangzhou Yuantiao Technology Co., Ltd. successfully financed and injected new vitality into the enterprise.

Compared with China Transinfo, Yuantiao Technology is undoubtedly the "younger generation" of big data transportation industry. However, the company was founded with its excellent strength in 2013 under the guidance of academicians of the Chinese Academy of Sciences. It has a high-end technical team composed of experts from the "Thousand Talents Plan", professors and doctors of Zhejiang University; thus, it has a strong innovation ability. In just a few years, it has made brilliant achievements in the field of the big data transportation. Its products include integrated traffic situation analysis machine and alarm machine its product system covers two major directions of "big data + traffic management" and "big data + transportation", and it has launched a special project of congestion control based on the big data analysis. The capital institutions see the potential of this enterprise, so the joint investment of Yinghua capital and SVB Capital China (Silicon Valley Bank) has successfully completed RMB 50 million Round A financing, which is used for the introduction of high-end talents and the construction of "big data + intelligent transportation" product system.

Similarly, in June 2017, another start-up also successfully financed. On June 20, Jiuqu Data completed the angel round financing of 2.25 million yuan invested by Shanghai Yisen. This enterprise, also based on big data transportation industry, focuses on big data real-time management and control system SVDM, providing customers with big data products such as data visualization and AI technology. The cooperative object of Jiuqu Data is the traffic management bureau. The system can collect static data such as vehicle model, displacement, and time of licensing, combine the driving behavior, location and other information collected by hardware, integrate the dynamic data such as weather and road conditions, analyze the regular running route of each vehicle so as to obtain the accident analysis index and display it in real time on the screen of traffic management command center. It is an active attempt of big data of transportation.

The smooth financing of two start-ups in succession shows that the attention of the big data transportation industry is constantly increasing. If we can come up with an excellent big data transportation scheme, which can directly solve the problems existing in transportation, then such start-ups are bound to be widely concerned by the capital market. With the successful financing of start-ups, the innovation of big data transportation industry will be further enhanced. It can be predicted that in the next few years China's transportation big data will show a trend of explosive development.

11.3.3 *Investment in big data transportation industry is surging*

While the start-up companies continue to win the favor of the capital market, large companies are also starting to make efforts in the field of big data transportation industry. Among them, there are many Internet giants like Baidu and Alibaba.

Baidu is undoubtedly the strongest. Baidu map and Baidu driverless car are Baidu Group's direct forays into the big data transportation industry. Subsequently, a series of investment and acquisition issues are also rapidly carried out. In November 2017, Baidu announced its strategic investment in Shouqi Limousine and Chauffeur. Baidu Capital, Weilai Capital, and Silu Huachuang took the lead, and completed a total of 700 million yuan of financing. Baidu has expressed its concern for big data transportation industry with practical actions, and has been making efforts in this field.

Alibaba's pace is also very firm. In October 2017, Alibaba Cloud launched cooperation with Shanghai Metro, developed technologies such as voice ticketing and face swiping, and announced to enter the field of big data intelligent transportation. Before that, Alibaba had accumulated rich experience in big data transportation industry through a series of investment and M&A. In 2013, Alibaba had a strong interest in the field of taxi, successively investing in fast, real-time bus service provider "Chelaile" and bus shuttle "Jie Wo Yun Ban Che". At the same time, Yitutong Technology and Amap are also Alibaba's investment brands.

Alibaba's big data technology is leading technology company in China. Alibaba cloud system has established cooperation with many commercial institutions, so it has high proficiency for big data application of transportation.

Tencent also has a rapid layout in the big data transportation industry. In December 2017, Zhonghe Technology announced to cooperate with Tencent, which entered the big data intelligent transportation market. Tencent's focus is on Zhonghe Technology's transportation industry experience. Based on Zhejiang University, Zhonghe is engaged in the construction of urban rail transit system and comprehensive ticketing system. It has already carried out business in metro projects in Shenyang, Xi'an, Zhengzhou, and other cities and has successively innovated and developed rail transit driverless signal system and intelligent train related products. It can be seen that Tencent is also covetous for the big data transportation industry.

Regardless of start-ups or Internet giants, more and more enterprises are participating in the big data transportation industry, which means that the outbreak of industry is coming. Transportation is related to people's basic necessities of life and reflects the development vitality and potential of a city, so we have reason to believe that in the future, the development of big data transportation industry will inevitably become more and more mature and lead the world trend.

11.4 Police Guidance of Big Data Transportation Industry

The development of the big data transportation in the business field is surging. The national policy field is also full of concerns about this, and constantly issues policy guidance to encourage the development of the industry.

11.4.1 *Definite support for the 13th Five-Year Plan*

In the 13th Five-Year Plan, the goal of building a modern comprehensive transportation system and realizing the development prospect of networking and intelligence is clearly put forward. The planning outline states that "we shall build a comprehensive transportation network to realize the integrated connection of transportation hubs. We shall gradually optimize the spatial layout of comprehensive transportation hubs and improve the construction of transportation hubs at the national, regional and local levels". In order to achieve this goal, we must actively access the big data system so as to form a data system mode and effectively connect all links involved in the transportation. This planning outline will further activate the vitality of the industry and lay a foundation for the future trend of the industry.

11.4.2 *The State Council's Guiding Opinions on Actively Promoting the "Internet +" Action*

In 2015, the State Council issued the "Guiding Opinions on Actively Promoting the 'Internet +' Action" and made clear explanations for the application of big data in the field of transportation. The Guiding Opinions state:

(1) **Innovation and development to support and lead.** We shall make full use of the Internet, big data, cloud computing, and other information technology to optimize the mode of transportation organization and provide diversified products to better meet the diversified needs. We shall lead by the development of intelligent transportation, enhance the innovation capacity of the industry, and foster new forms and models of business development.
(2) **Market operation to improve quality and efficiency.** We shall give full play to the initiative of traditional transportation enterprises and Internet enterprises, and encourage all-round cooperation in the form of capital operation, technical cooperation, management cooperation, etc. We shall give full play to the advantages of technology and market, focus on customers and improve the overall operation efficiency and service quality of the comprehensive transportation system.

The "Internet +" convenient transportation key demonstration project shall be implemented by 2018 and the public getting access to the dynamic road traffic information in the mobile Internet terminal and completing the one-stop service of the entire passenger transportation including navigation, ticketing and payment shall be basically realized so as to enhance the user's travel experience. The interconnection and interworking of "T-union" in key urban agglomerations and "one in one network control" for key operating vehicles (ships) shall be basically realized. Online and offline enterprises speeding up integration and taking the lead in realizing "one order to the end" in the national backbone logistics channel shall also be aimed at. The Internet of transportation infrastructure, transportation tools, and operation information shall be realized to make the system operation safer and more efficient.

It shall focus on improving traffic capacity, easing traffic congestion, reducing emissions, and improving the coverage rate of ETC lanes of national expressways. The convenience of ETC system installation and payment shall be improved. The development of users shall increase. Improving the utilization rate of road, vehicles, taxis, and other types of operating vehicles shall be focused and achieving more than 50% of the utilization rate of bus ETC within three years shall be strived. The standard van type freight ETC shall be researched and promoted. The service level of customer service outlets and provincial online settlement centers shall be improved, and an efficient settlement system shall be built. The deep integration of ETC system and Internet shall be promoted to realize the extensive application of ETC system in the transportation fields such as highway, urban public transportation, taxi, parking, road passenger transport, and railway passenger transportation.

The respective advantages of transportation enterprises and Internet enterprises shall be given full play and the integration of online and offline resources shall be encouraged to provide diversified and high-quality services for the public. The potential needs of passengers shall be explored and met, and the offline resources shall be relied to expand online, extend service chain, and innovate business model. The characteristics of the Internet, such as personalized demand, timely response, and efficient organization shall be given full play. The offline resources shall be actively integrated and the "Internet +" transportation new formats in the fields of urban transportation, road passenger transportation, freight transportation, parking, and vehicle maintenance shall be developed through customized transportation, network booking of taxis, and time sharing leases to gradually realize the scale, networking and branding to promote public entrepreneurship and innovation. The transportation enterprises and Internet enterprises shall be encouraged to carry out strategic cooperation to achieve deep integration of information resources, capital, technology and business and combine with upstream and downstream industrial chains.

11.4.3 *Notice of the Ministry of Transport*

In June 2015, the Ministry of Transport issued the Notice of the General Office of the Ministry of Transport on Further Accelerating the

Construction of Intelligent Application Demonstration Projects of Urban Public Transportation, further promoting the development of big data intelligent transportation and made clear requirements:

In order to ensure the orderly construction of intelligent application demonstration projects of urban public transportation, different types of cities are promoted in a step-by-step and phased manner.

(1) By the end of 2015, the construction of main body of the demonstration projects in the first batch of 10 pilot cities will be completed. The provincial transportation authorities in Jinan, Zhengzhou, Dalian, Harbin, Shenzhen, Nanjing, Xi'an, and Changsha and the transportation authorities in Beijing and Chongqing, which have applied for the first batch of construction of demonstration projects, should take the city as the main body, further accelerate the construction progress of the demonstration projects, complete the construction of the main body of the demonstration projects by the end of 2015, including the public transportation data resource center, the intelligent dispatching platform for enterprise operation, the industry supervision platform and the passenger travel information service platform, and at the same time ensure that the construction of the demonstration projects conform to the relevant standards and specifications.

(2) By the end of 2016, the construction of main body of the demonstration projects in the second batch of 27 pilot cities will be completed. The provincial transportation authorities in Taiyuan, Shijiazhuang, Qingdao, Wuhan, Zhuzhou, Guiyang, Suzhou, Urumqi, Hangzhou, Baoding, Yinchuan, Lanzhou, Kunming, Ningbo, Hefei, Nanchang, Xinxiang, Guangzhou, Shenyang, Xining, Liuzhou, Fuzhou, Haikou, Hohhot and Changchun and the transportation authorities in Shanghai, and Tianjin, which have applied for the second batch of construction of demonstration projects, should take the city as the main body, speed up the start of the construction of the demonstration projects and complete the construction of main body of the demonstration projects by the end of 2016. For the pilot cities that have not completed the preliminary work of the demonstration projects the preparation and approval of the project feasibility research report, fund application report, and preliminary design shall be accelerated to complete and reported to the Ministry of Transport in time in accordance with the ***Notice of the General Office of the Ministry of Transport on Regulating the Preliminary Work Management of the***

Industry Informatization Construction Projects (GMTP [2014] No. 41) and the requirements of relevant documents.

(3) By the end of June 2017, the construction of demonstration projects in 37 demonstration cities will be completed.

11.4.4 *Implementation Plan of Intelligent Transportation Development*

In 2016, the National Development and Reform Commission and the Ministry of Transport issued the *Implementation Plan on Promoting the Internet + Convenient Transportation to Promote the Development of Intelligent Transportation*:

(1) The importance of promoting the Internet + convenient transportation and promoting the development of intelligent transportation shall be fully understood. The convenient travel of passengers and efficient transportation of goods shall be taken as guidance to comprehensively promote the wider and deeper integration of transportation and Internet so as to provide strong support for the modernization of China's transportation development.
(2) Seriously organize and conduct the Implementation Plan. Local governments and departments at all levels should strengthen organizational leadership, clarify the division of responsibilities, take the initiative in combination with the actual situation, actively promote the implementation of key demonstration projects and focus on the implementation of tasks.
(3) Increase capital investment. The leading role of government investment in demonstration and leverage shall be given full play to fully attract social capital to participate in the construction and operation of intelligent transportation. An open and inclusive development environment shall be built. The laws and regulations shall be improved and the standardization of new formats and models shall be encouraged.
(4) The Development and Reform Commission and the Ministry of Transport will organize and implement relevant work as a whole, strengthen coordination and guidance, carry out supervision and promotion of key demonstration projects and effect evaluation and summarize the promotion experience in time.

Chapter 12

Unmanned Driving: Transportation Revolution Sweeping Across the Globe

Unmanned driving set off a transportation revolution sweeping the world, Starting the New Era of Intelligent Transportation. The autonomous driving is the main trend for the development of Transportation, and disrupting and Reshaping Transportation Ecology in the long run. Driverless taxis and shared travel, driverless truck fleets will become commonplace. In cities, rail transit will also be automated and driverless. Driverless taxis and shared travel, driverless truck fleets will become commonplace. In cities, rail transit will also be automated and driverless. Driverless taxis and shared travel, driverless truck fleet will become widespread, inside the city, rail transit will be automated and unmanned. At the same time, our roads will become safer and more efficient.

12.1 Origin, Development, and Application of Unmanned Driving Technology

12.1.1 *Unmanned driving starting the new era of intelligent transportation*

Promoted by the thriving scientific technology, the role of the unmanned driving technology in social studies is increasingly important. Based on the surrounding-perceiving technology, the unmanned driving automobile can perceive the surroundings around the automobile and process the information about the surroundings. Unmanned driving technology can guarantee the automobile's normal working on the road and its reaching its planned destination by making use of computer information technology to control and adjust the automobile's speed and driving direction.

During the process of the development of the unmanned driving automobile, accurate information technology has been applied to the automobile to develop the people–automobile interaction, whose efficiency has been increased while the operating complexity has been decreased. The unmanned driving technology may not only increase the automobile's operating efficiency but also avoid the chances of traffic accidents and offer travel service for those who have not learned to drive.

12.1.1.1 *Profile of development in the unmanned driving technology*

In the long run, autonomous driving is the main trend for the development of automobiles and unmanned driving is a type of display of autonomous driving and an important component of intelligent transportation. Broadly speaking, the unmanned driving automobile can be regarded as the integrated application of computer technology, communication technology in a network, and technology of intelligent control in the context of the internet. The unmanned driving automobile is a sort of special intelligent robot which can move, thus many experts in this field also call it the moving robot with automatic wheels.

The unmanned driving automobile can use on-board sensors to achieve automatic perception of the surrounding environment, grasp real-time road condition information, and through the intelligent computer system as the core of the intelligent driving device, the automobile can be driven to the destination without any intervention. During the driving process, the unmanned driving automobile can use the system equipped with safety control functions to deal with a variety of unexpected situations to ensure the safety of driving.

12.1.1.2 *Direction of applying the unmanned driving technology*

(1) The Unmanned Driving Systems in the Highway Environment: Highways are well-marked with road signs and the unmanned driving systems used in this structured environment need to assume the function of tracking road markings and automatically identifying vehicles. For high-speed automated driving in this more standardized

environment, the unmanned driving systems strive to achieve a fully automated level of driving. These applications are difficult to apply to other environments, but the full value of unmanned driving applications can be demonstrated if fully automated driving can be achieved because highway driving is so dangerous and boring.

(2) The Unmanned Driving Systems in Urban Environments: Compared to driving on highways, the unmanned driving in the city has a broad prospect for development as it requires less automobile speed and it is somewhat safer. At present, it can share the pressure of high-capacity public transportation in the city and relieve the traffic tension in urban areas. The specific application scenarios include airports, industrial parks, parks, campuses, or other public places.

(3) The Unmanned Driving Systems in Special Environments: Some countries have more advanced technologies for the unmanned driving and have committed to the use of unmanned technology in military and other special environments. There are commonalities among the unmanned driving technology in these areas and applications in highway and urban environments, but there are differences in the relevance of technical performance as well.

12.1.1.3 *Bottlenecks in the development of unmanned driving technology*

The unmanned driving automobiles are mostly at the experimental and testing stage, which have a long way to go before they can be put into use in the real complex traffic environment, and there are various bottlenecks in their development: laws and regulations, technical limitations, data leakage, ethics, and so on. In the near future, the unmanned driving automobiles are unlikely to think and make decisions like humans, and the protection of the safety of the vehicle and the passengers inside is the primary task of the unmanned driving automobile. In the manned driving situation, most drivers will sacrifice some of their own interests in order to protect the lives of others if their life is not endangered.

For example, if the pedestrian is running a red light at the front of the road, the other vehicles are traveling at the left side in the same direction and a roadblock is on the right side, in this case, many drivers would prefer to turn right into the roadblock in order not to endanger others' lives even if the pedestrian is running the red light though repairing the broken car will cost them some money. The automatic driving automobile, however, will judge the front resistance as the minimum so as to directly hit the pedestrian. Therefore, the journey of the development of the unmanned driving technology is not smooth, which requires not only in-depth layout of the government, universities, research institutions, but also the majority of entrepreneurs' and enterprises' active participation with the strong force of the market to speed up the process of its development so that the subversive unmanned driving technology and intelligent transportation can really benefit hundreds of millions of people.

12.1.1.4 *Future outlook for the unmanned driving technology*

After the birth of the automobile, the progress of the automobile industry has made a positive contribution to the development of society. Relying on advanced technological means, the unmanned driving effectively improves the safety and accuracy of driving and reduces the chances of traffic accidents involving automobile driving. Along with the further development and maturity of the technology, the scope of application of the unmanned driving in passenger automobile and commercial vehicles will also continue to broaden, which will set off a revolutionary wave on the global scale.

China's huge population has laid a sound market foundation for the development of the automobile industry, which occupies a very important position in the global automobile development. In recent years, the domestic automobile industry has been focusing on building its own brand of products, creating a good opportunity for the application of new automobile technology, which is conducive to expanding the scope of application of this new automobile technology in the field of automobiles in China. The intelligent driving may drive the development of the entire automobile field and promote the progress of commercial vehicles. Analyzed from the perspective of passenger automobiles, the commercial vehicles using the intelligent driving technology can serve in people's daily life and work, save them more time and energy, and thus improve their quality of lives.

For example, if the commercial vehicle on the highway can make unmanned driving a reality, it will give the drivers more time to interact with other people about their current location and the estimated time of arrival. If unmanned driving is used in the logistics industry, it will improve the distribution efficiency of the logistics center to speed up the operation of the whole system. For commercial vehicles, the application of the unmanned driving technology can effectively provide more time for the drivers to rest, to avoid driving fatigue, reduce the chances of traffic accidents, and avoid endangering people's lives and property safety.

12.1.2 *Four stages of the development of the unmanned driving technology*

In response to the explosive growth of active safety technology in automobiles, the National Highway Traffic Safety Administration (NHTSA) released a five-level standard for the automation of automobiles in 2013, which classifies the automated driving function of vehicles into five levels: Level 0 to Level 4.

12.1.2.1 *Level 0: No automation*

Level 0 means that the automobile does not have any autopilot function and the driver has full control over all the functions of the automobile. While driving the car, the driver is responsible for the starting, braking, and operating of the automobile and observing the surrounding road conditions as well. As long as the human controls the automobile, it is classified as Level 0 no matter what driving assistance technology the automobile has been equipped with. Therefore, although the regular car is equipped with functions such as forward collision warning, lane departure warning, automatic wiper control, and automatic headlight control, it can only be driven with human control and it is still classified as Level 0.

12.1.2.2 *Level 1: Automation at single-function level*

Level 1 means that the driver is responsible for the driving safety, but the driver can relinquish some of the control and leave it to the system. At this stage, some of the functions have been already automated, such as the frequent adaptive cruise control, emergency brake assistance, lane keeping, and so on. Level 1 has only a single function and the driver is still in control of the automobile while driving and the automobile cannot be left completely alone.

12.1.2.3 *Level 2: Partial automation*

At Level 2, the automobile can share the rights of control with the driver. In some of the predefined environments, the driver does not need to control the automobile and the automobile can work automatically, but the driver is always on standby, responsible for the driving safety and ready to drive the vehicle at any time. At this stage, the key to partial automation of the automobile is not to have more than two functions, but that the driver is no longer the primary operator of the automobile. For example, the tracking function formed by the combination of ACC and LKS and the autopilot function invented by Tesla are of Level 2 functions.

12.1.2.4 *Level 3: Conditional automation*

At this stage, the automobile can be controlled automatically in limited situations, such as driving on highways or roads with little traffic. The driver has enough time to take over the car in case of an emergency to ensure safe driving. At this stage, the driver is freed up to the greatest extent possible and no longer has to be responsible for the driving safety and constantly monitoring road conditions while driving.

12.1.2.5 *Level 4: Full automation (unmanned driving)*

At this stage, the automobile can travel to its destination without human assistance by entering the departure and arrival information in advance. The automobile will be responsible for the driving safety throughout the entire travel with minimal reliance on the driver, and the vehicle can travel without a passenger.

In addition to the NHTSA, the Society of Motor Vehicle Engineers (SAE) has also graded automated driving, dividing automated driving technology into six levels from 0 to 5, with levels 0 to 3 being consistent with the NHTSA's classification of automated driving, which respectively emphasizes no automation, driving support, partial automation, and conditional automation. As for full automation, SAE has subdivided it and further emphasized the requirements for driving on the road and in the environment.

At this stage, according to SAE's regulations of Level 4 for automated driving, the automobile can only be automated in certain road conditions such as closed parks and fixed driving routes. Level 5, on the other hand, allows for automated driving in a variety of environments and provides effective responses to complex vehicle, pedestrian, and road conditions.

All in all, the driving functions at different levels increase accordingly. Advanced Driving Assistance Systems (ADAS) is the highest level of driver assistance system and belongs to Level 0 to Level 2 (see Table 12.1).

Among them, Level 0 autopilot has only sensor detection and decision alert functions; if it is equipped with a night vision system, it will be able

Table 12.1 Automatic Driving Functions Increasing from L0 to L5

NHTSA	L0	L1	L2	L3	L4	
SAE	L0	L1	L2	L3	L4	L5
	No automation	Driving assistance	Partial automation	Conditional automation	High automation	Full automation
Functions	Night vision, pedestrian detection, traffic sign recognition, blind spot detection, road merging assistance, rear crossing alert, lane departure warning	Adaptive cruise control, automatic emergency braking, parking assistance, forward-back collision warning system, electronic stability system	Lane keeping assistance	Congestion driving assistance	Auto parking in the parking lot	—
Features	Sensory detection and decision alerts	Single function (one of the above)	Combined functions (combination of L1/L2)	Specific condition, specific task	All condition, all task	—
		ADSA			Automatic driving	

to detect pedestrians in front of it, recognize traffic signs, and issue warnings when the vehicle deviates from the lane and so on.

- The automatic driving at Level 1 has a single control function, such as active emergency braking, adaptive cruise control, and so on.
- The automatic driving at Level 2 can perform multiple control functions, such as configuring the vehicle with AEB and LKA.
- The automatic driving at Level 3 can be performed under specific conditions and the control of the automobile needs to be handed over to a human driver if it goes beyond the limited conditions.
- Level 4 defined by SAE allows for unmanned driving under specified conditions, such as unmanned driving on a fixed route.
- Level 5 defined by SAE refers to complete unmanned driving, which is the highest level and the final form of automated driving, allowing the unmanned driving in any scenario.

Compared with the semi-automated ones, the fully automated unmanned driving automobiles are safer because the former can be subject to human errors while the fully automated unmanned driving ones cannot be. For example, a survey by the Virginia Tech Transportation Institute (VTTI) showed that it takes 17 seconds for a driver of a Level 3 automated automobile to respond to takeover request from the automobile, and if the automobile is traveling at 65 mph, the vehicle will have covered 1,621 feet, or 494 meters, during these 17 seconds.

From seeing an object on the road to applying the brakes to the automobile, it will take the human driver 1.2 seconds, which is much longer than the 0.2 seconds used by the on-board computer. Although the gap is only 1 second, the automobile can travel 40 meters in 1.2 seconds if the automobile is traveling at 120 km/h, and it can only travel 6.7 meters in 0.2 seconds. In the case of an emergency, this one second gap could be the very one between life and death for the passengers. Therefore, for the entire automotive industry, unmanned driving is the ultimate goal of automated driving.

12.1.3 *The evolutionary path of unmanned driving technology*

The unmanned driving technology is a comprehensive technology based on the traditional automobile manufacturing technology which integrates the technology of intelligence, automation, electrification, Internet, and other technologies. The unmanned driving automobile, also known as an "automated vehicle" or "wheeled mobile robot", perceives the nearby road environment through the vehicle's sensor system, makes reasonable planning for the driving route, and controls the vehicle's progress toward the predetermined destination. The unmanned driving automobile is the product of the development of various technologies such as automatic control, artificial intelligence, pattern recognition, and visual computation and it is also the product of the technology of computer science, pattern recognition, and intelligent control technology to a certain stage, and it is now regarded as an important indicator of a country's scientific research strength and economic development level, which can be widely used in national defense, national economy, and other fields in the future.

China has a large population and the traffic problems in the country are very serious ones. The urban traffic troubles, environmental issues, and urban development problems will be solved, the level of industrial manufacturing will be effectively enhanced and people's lives will be improved as soon as possible if unmanned driving can be fully promoted. With the gradual maturation of the unmanned driving technology, the traffic management departments must be prepared to promote the integration of the unmanned driving technology into intelligent transportation to achieve comprehensive development.

In recent years, both the domestic and foreign high-tech enterprises as well as the traditional automobile manufacturers have actively deployed efforts in the field of the unmanned driving, giving birth to many products, many of which have passed the road tests, and some of the tests have covered hundreds of thousands of kilometers, which has laid the sound foundation for eventual commercial mass production of the unmanned driving automobiles. It is said the intelligent networked automobile will become the most promising industry in the future. An analysis of the plans released by the major automobile manufacturers and Internet companies revealed that the first year of the commercialization of the unmanned driving automobile, after which the commercial application of the unmanned driving automobile would enter an explosive growth stage.

From an overall point of view, there are two developing paths for the unmanned driving automobile, one of which is the ADAS and the other is artificial intelligence. The highest form of both the ADAS and the artificial intelligence agree with each other and both support fully unmanned driving, that is to say, fully unmanned driving is the result of automation and intelligence development to the highest stage.

In the future, the automotive industry is bound to continue to develop in the direction of the intelligent automobile, and eventually it will emerge as the unmanned driving automobile. It is the unmanned driving automobile that shoulders the responsibility of reducing traffic accidents, easing traffic congestion, and reducing environmental pollution, but there are also problems with safe driving, popularity of recognition, legal ethics, and other aspects as well. According to experts' predictions, unmanned driving would become a reality, and by 2030, the value of the unmanned driving market will reach $87 billion, and by then, what will the applications of the unmanned driving technology achieve?

The final goal for the intelligent automobile is to make unmanned driving a reality. The application of unmanned driving technology may meet three major potential needs, which are perceivability, connectability, and the personalization of the automobile. The achievement of the universal application of the unmanned driving technology requires us to create an intelligent automobile standard system and intelligent transportation system, and eventually make automobile sharing, easing of traffic congestion, and reduction of environmental pollution a reality so that the full liberation of human and automobile productivity can be achieved.

12.1.3.1 *Application of the unmanned driving technology*

The application of unmanned driving technology should address three major potential needs: the first one is perceived need from the real world to the digital world, which is the need that can be understood by people and devices through structured data including the need for people to perceive the automobile, the need for the automobile to perceive people, and the need for the automobile to perceive the surrounding environment. The second is the need for the connection among different devices, such as smartphones, smart wearables, smart automobiles, and so on. The third is the demand for the personification of the automobile, and the automobile in the future may develop into an intelligent robot that can not only act on instructions from people, but also guess their needs and recommend life services for them.

12.1.3.2 *Promotion of the unmanned driving technology*

It is essential for us first to establish and improve intelligent automobile standards, intelligent automobile laws and regulations, intelligent transportation systems, and then enhance the cooperation of the information industry, electronics industry, transportation industry, the Internet industry, and other industries concerned to promote unmanned driving technology, giving full consideration to the future development trend of the technology, hardware and software interfaces, data communication formats and protocols, platform security construction, and international intelligent automobile technology standards, and then the intelligent automobile standard system with Chinese characteristics will be established. In addition, China will also speed up the construction and implementation of intelligent automobile pilot demonstration projects, build an intelligent traffic network, issue intelligent car-related laws and regulations, and make the issues such as the determination of responsibility after a traffic accident involving an automated automobile clear.

12.1.3.3 *Mission of the unmanned driving technology*

With the support of the unmanned driving technology, the shared automobile will continue to develop, and the automobile will serve the dual functions of public transport and services. With the increasing popularity of the shared car, the demand for the purchase of private cars will continue to decline, traffic congestion and parking difficulties will be effectively alleviated, and greenhouse gas emissions will be significantly reduced. Coupled with the time limitation, safe and accurate operation of the shared automobile, the unmanned driving technology will not only reduce the frequency of traffic accidents, but also allow the disabled, people without driver's licenses, people without cars, and the elderly to enjoy the fun of driving, which will fully liberate people's time and productivity.

12.1.4 *The unmanned driving disrupting and reshaping the transportation ecology*

The ecology refers to the living conditions of all living things and the interlocking relationships between them, and between them and their environment. Accordingly, the transportation ecology refers to the relationship between vehicles, roads, and other elements within the transportation system, as well as between them and the external environment. At present, there are many problems with the transportation ecology in China such as frequent traffic accidents, serious traffic congestion in cities, and severe air pollution. The emergence of the unmanned driving technology has brought new ideas and solutions for these problems.

It had been predicted that by 2021, the unmanned driving automobile will fully enter the market, overturning the existing pattern of the automobile industry and promoting the transformation and upgradation of the automobile industry into a new stage. According to data released by the World Economic Forum, the value created by the digital transformation of the automotive industry will exceed $67 billion, and the social benefits created will reach $31 thousand ($3.1 trillion), including the improvement of unmanned vehicles, passenger connectivity, and improvements of the entire ecosystem of the transportation industry.

It is estimated that the semi-autonomous and fully automated vehicles have great market growth potential in the coming decades. For example, by the year of 2035, the number of automated vehicles in China will reach 8.6 million, of which 3.4 million will be fully automated unmanned driving vehicles and 5.2 million will be semi-automated vehicles.

According to the prediction of the industry authorities concerned, China's car sales, buses, taxis, and related transportation services could generate about $1.5 trillion in revenue annually. According to the Boston Consulting Group's forecast, it will take at least 15 to 20 years for the unmanned vehicles to reach a 25% share of the global automobile market. In other words, if the unmanned vehicles go on sale in 2021, it means that it will take until 2035–2040 for their share of the global car market to reach 25%.

It is because of the unprecedented impact that the unmanned vehicles have had that they are causing a huge change in the automobile industry. Related studies show that with the advent of the unmanned driving automobile, the safety on the highway will be greatly enhanced and air pollution and traffic congestion will be effectively alleviated.

12.1.4.1 *Enhancing highway safety*

Today, highway accidents are a major problem troubling the world. According to the statistics, in the United States each year about 35,000 people die in car accidents, in China each year about 260,000 people die in car accidents, and in Japan each year about 4,000 people die in highway accidents. According to the statistics from the World Health Organization, the number of people killed in highway accidents around the world each year is as high as 1.24 million.

According to the statistics, the annual loss due to fatal car accidents is up to $260 billion, the loss in the injuries caused by car accidents is up to $356 billion, and the annual loss due to highway accidents is about $625 billion. According to the survey by RAND Corporation, among various factors causing the fatal traffic accidents, drunken driving accounts for 39%. In this regard, the widespread application of the unmanned driving vehicles will effectively alter this situation and reduce or even avoid car accident casualties. In addition, according to the data released by the transportation department, 60% of all traffic accidents in China are caused by the collisions between pedestrians, bicycles and electric bicycles, and cars and trucks. And the human error is responsible for about 94% of all motor vehicle accidents in the United States.

Research by the Insurance Institute for Highway Safety in the United States shows that with the installation of automatic safety devices (collision braking system, collision warning system, lane departure warning, blind spot detection, etc.), cars will be safer on the highway and the highway accident fatalities can be reduced by 31%, that is, about 11,000 people will be prevented from being lost in traffic accidents per year.

12.1.4.2 *Relief of traffic congestion*

Traffic congestion is an epidemic in major cities. In the United States, each driver wastes about 40 hours a year to be stranded on the road due to traffic congestion, and the cost of traffic congestion is about $121 billion. In the major cities such as Moscow and Rio de Janeiro, the time wasted in traffic congestion is even worse, with each driver wasting more than 100 hours per year in traffic jams. In China, the number of the cities which have more than 1 million cars is about 35 and the number of cities which have more than 2 million cars is 10, and about 75% of the roads in urban areas of large cities are congested. The number of private vehicles has now reached 126 million, a 15% increase from the previous year.

Donald Shoup's study found that 30% of traffic congestion in urban areas in large cities is caused by drivers circling back and forth looking for parking spaces, and this is the very factor that causes traffic congestion and air pollution, as 30% of carbon dioxide emissions is from cars, and carbon dioxide is considered responsible for climate change. In addition, it is estimated that 23–45% of traffic congestion occurs at road intersections. Because traffic lights and stop signs are stationary and the time intervals are set in advance, they cannot take the traffic flow into account in all directions, making it difficult for them to play a real role in regulation of the traffic flow.

As the unmanned vehicles account for an increasing proportion of the traffic flow, the in-vehicle sensors will be combined with the intelligent transportation systems to optimize the traffic flow at the road intersections, and the time intervals of traffic lights will be automatically adjusted according to the traffic flow in each direction, so that the efficiency of vehicle traffic flow at road intersections will be improved to provide an effective solution to the traffic congestion problems.

12.1.4.3 *Reducing air pollution*

Cars are a very important contributor to air pollution. According to a study by RAND (James M. Anderson, Nidhi Kalra, Karlyn D. Stanley, Paul Sorensen, Constantine Samaras, Tobi A. Oluwatola etc. www.rand.org/t/RR443-2), unmanned driving technology may improve fuel efficiency by 4% to 10% through smoother acceleration and deceleration compared with manual driving. According to the survey, smog in industrial areas is closely related to the number of cars, and some studies have proved that the exhaust emissions from cars forced to stop at red lights or in traffic jams are 40% higher than those emitted during normal driving. The increase in the number of unmanned vehicles can alleviate air pollution to a large extent.

The unmanned driving vehicle sharing systems may help to achieve energy savings. Researchers at the University of Texas at Austin did a study of carbon monoxide, sulfur dioxide, nitrogen oxides, volatile organic compounds, fine particulate matter, and greenhouse gases and found that the unmanned car-sharing systems can save energy and reduce emissions of many types of pollutants.

Uber found that 30% of the passengers in Los Angeles and 50% of passengers in San Francisco choose to carpool. Globally, about 20% of the passengers choose to carpool. Whether unmanned or conventional cars, the more the passengers choose to carpool, the better the environmental benefits and the more the traffic congestion will be alleviated. The air quality is effectively improved because carpooling changes the pattern of one-person-one-car rides.

12.2 Global Distribution: Competing for the High Ground of Unmanned Driving

12.2.1 *Competitive pattern of the global unmanned driving industry*

The unmanned driving automobile is a type of smart car, also known as the wheeled mobile robot, which makes unmanned driving a reality by relying mainly on the smart driver in the car.

In simple terms, the unmanned driving automobiles make the automatic driving of the vehicle a reality on the basis of high-precision maps, which are supplemented by the data collected by onboard sensing devices and successfully work with the help of intelligent algorithms to identify, and operations to make decisions, to control the vehicle. As can be seen through the working principles of the unmanned driving vehicles, the unmanned driving technology is an integration of technological achievements in various fields such as artificial intelligence, vehicle braking, environmental recognition, etc., bringing new development opportunities for the automotive, smart chips, sensors, and map navigation markets.

12.2.1.1 *A look at the unmanned driving industry from California road test qualifications*

Currently, California is the first in the world to pass the unmanned driving legislation and it is the premier test site for the unmanned driving automobile in the world. The US. Highway Traffic Safety Administration, which oversees US auto safety, is headquartered in California, and because of its openness, authority, and inclusiveness, California has become the world's leading test site for unmanned driving cars.

The Department of Motor Vehicles (DMV) in California issues California Driverless Road Test Permits for the unmanned driving automobile companies, which are granted to test unmanned driving automobiles on specific public roads. In September 2014, the DMV issued the first driverless road test permits (see Figure 12.1). To date, 45 companies have been granted the driverless road test permits all over the world, and these companies cover a diverse group of enterprises including the traditional car companies, parts suppliers, technology giants, and startups.

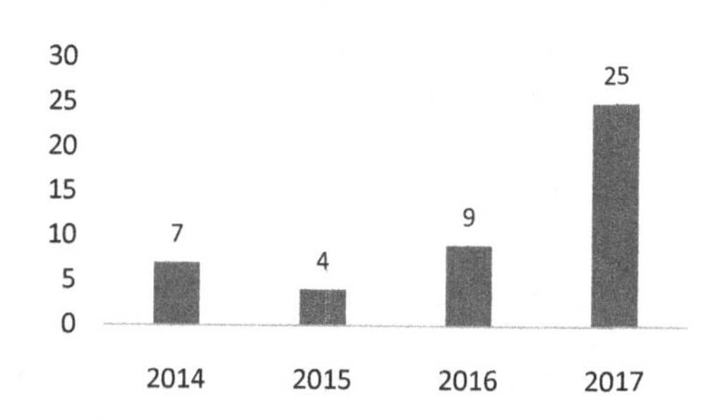

Figure 12.1 Accelerating the grant of California driverless road test permits.

The global unmanned driving industry is growing at an increasingly rapid pace. In 2014, seven companies were granted California's driverless road test permit, and four companies were granted the permit in 2015, nine in 2016, and 25 in 2017. The percentage of companies that was granted California's driverless road test permit in 2017 alone was more than half of the total, showing the explosive growth trend. This phenomenon shows that the unmanned driving industry is growing at an increasingly rapid pace, with major companies around the world focusing on the study of the unmanned driving automobile, and the competition in this field is becoming increasingly intense.

With the release of the Smart Grid Roadmap, the core technology of the unmanned driving automobile will be further defined. With the intelligent network connectivity roadmap, the core technologies required for the unmanned driving will be further clarified, and the entire industry will move toward network connectivity and other core unmanned driving technologies.

The Unmanned Driving Industry Development Prospects and Investment Strategic Planning Analysis Report in China released by the Foresight Industry Research Institute shows that by 2025, the market value of the unmanned driving will reach $200 billion to $1.9 trillion. By 2035, the global sales of the unmanned driving automobile will reach 11.8 million units. From 2025 to 2035, the compound annual growth rate of the unmanned driving automobile will reach 48.35%, by then China's share of the global unmanned driving car market will reach 24%.

12.2.1.2 *The penetration rate of global automated driving will increase rapidly, and the market space may exceed hundreds of billions of dollars*

SAE divides automated driving into six levels, with Level 0 being completely manual driving, Level 5 being completely unmanned driving, and the intermediate being varying degrees of manual driving and automated driving. At present, the automatic driving technology at Level 1 and Level 2 has been relatively mature, the automatic driving technology at Level 3 will achieve mass production, while the completely unmanned driving at Level 5 will take more than ten years to achieve industrialization. It is estimated that in the near future, the global penetration rate of automated driving will increase rapidly. It was expected that by the year of 2020, the penetration rate of Level 1 and Level 2 automated driving

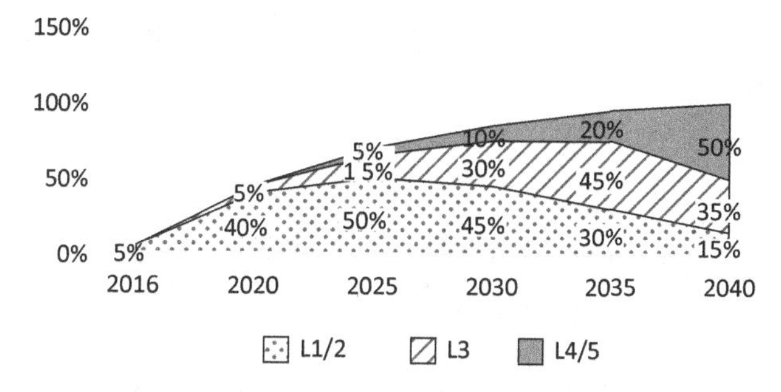

Figure 12.2 Penetrating ratio of unmanned driving cars at different levels.

Table 12.2 Issuing Time for California Road Test Permit

Issuing Time	Companies
2014	Volkswagen, Mercedes-Benz, Google Waymo, Delphi, Tesla, Bosch, Nissan
2015	GM, BMW, Honda, Ford
2016	Zoox, Drive.al, Faraday Future, Baidu, Wheelgo, Valeo, NIO, Telenav, Nvidia
2017	AutoX, Subaru, Udacity, Navya, Renovo, Uber, PlusAl, Nuro, CarOne, Apple, Bauer, Xiaoma Zhixing, Tucson Future, Jingchi, SAIC, Almotive, Nullmax, Samsung, Voyage, Continental, CYNGN, Roadster, Changan, Lyft

will reach 40%; by the year of 2025, more than 20% of mass-produced cars are expected to achieve different levels of automated driving; by 2040, all newly produced cars will be equipped with automated driving functions, of which the penetration rate of Level 4 and Level 5 automated driving cars will reach 50%, and the corresponding market size will exceed hundreds of billions of dollars (see Figure 12.2).

12.2.1.3 *Startups are an important force in the global unmanned driving industry promoting multi-industry integration*

By the end of 2017, 45 companies had been granted unmanned driving permits in California, among which 11 were traditional automobile manufacturers, such as Volkswagen, GM, Changan, etc., six were automobile parts suppliers including Delphi, Valeo, Continental, Bosch, etc., seven were tech giants, such as Apple, Tesla, NVIDIA, etc., and 21 were startups, such as Zoox, Drive.ai, Celeste, Pony.ai, etc. (see Table 12.2),

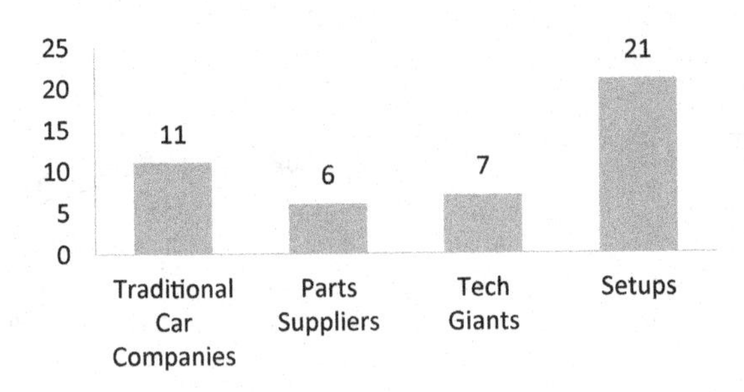

Figure 12.3 Startups account for the largest share.

of which the startups account for 47%, or nearly half of the total, making them a significant participator in the global unmanned industry (see Figure 12.3).

Accordingly, we may judge that the future of automotive technology may develop in the direction of the deep integration of machinery, electronics, communication, and artificial intelligence. Traditional car companies, famous universities, and technology giants will give birth to more startup teams, which are expected to achieve rapid development because of their crossover backgrounds.

12.2.1.4 *High-speed development of startups reshaping competitive landscape of the unmanned driving industry*

In 2014 and 2015, none of the startups were granted California unmanned driving licenses; in 2016, five companies were granted California unmanned driving licenses, accounting for 56% of the total; and in 2017, 16 companies were granted California unmanned driving licenses, accounting for 64% of the total, which shows a rapid development trend in this field. The startups are expected to grow by leaps and bounds as they are just entering the industry and are lightly loaded. Accordingly, we may judge that the startups are most likely to take important positions in the manufacturing and operations of the unmanned vehicles, system solutions, key components, and other fields by virtue of their advantages.

12.2.2 *Development of the unmanned driving technology around the world*

The continuous development of transportation technology has brought a lot of convenience to people's travel, and the speed of the car updates and iterations are getting faster and faster. In addition to the strong power, personalized design, and good human–vehicle interaction experience required, it also puts forward higher requirements in terms of environment protection, safety and intelligence, etc. In this context, the unmanned driving technology has a broad application prospect and is the mainstream trend of the automotive industry.

12.2.2.1 *Development of the unmanned driving technology in foreign countries*

The research on the unmanned driving cars in Western developed countries began in the 1970s, and major breakthroughs have been made in the feasibility and practicality. By the late 1980s and early 1990s, many countries detached from unmanned driving automobile research projects and began to focus on highway civilian vehicle-assisted driving projects because the expected goals were too complex and the technical conditions were immature. The unmanned driving cars of this period, however, were semi-autonomous vehicles, and there was a large gap between them and the real unmanned driving cars.

Until 1999, the unmanned driving car developed by Carnegie Mellon University in the United States successfully crossed the east and west of parts of the United States. On 5,000 km of the intercontinental highway, the vehicle traveled more than 96% of the distance at a speed of 50 to 60 km/h autonomously, thus the unmanned driving car project was declared a success.

In 2000, Toyota Motor Corporation had developed the unmanned driving bus and the road guidance, tailgating prevention, fleet driving, operation management, and other parts consisted of the automatic driving system. In 2007, Germany launched the unmanned driving car Lux and this unmanned driving car was transformed from a normal car which made use of laser sensor technology, intelligent computers, and global positioning devices. The unmanned driving Lux could deal with complex road conditions and drove automatically in the urban highway system.

In 2010, a team led by Professor Rogas from the Free University of Berlin modified a Volkswagen Passat to develop an unmanned driving car that could accurately identify other people and vehicles on the road, even in underground tunnels. This unmanned driving car could adjust the driving status of the car with traffic lights by using cameras, laser scanners, heat sensors, satellite navigation, and other technologies and devices.

In order to adopt more scientific and reasonable driving strategies in response to the road conditions, the unmanned driving cars usually need to build 3D images in the car's computer. The unmanned driving car may formulate a scientific and reasonable driving route for the car to provide passengers or controllers with an estimated time of arrival through the analysis of road conditions, vehicle position, and other data.

In 2010, Heathrow Airport in London made use of unmanned driving car called Uterra, which was jointly developed by the University of Bristol and some advanced companies in the transport system, but the car needs to work on a dedicated road.

The unmanned driving Saikabo developed by the French company INRIA is relatively mature, applies the advanced GPS global positioning technology, and is equipped with a touch screen, which allows passengers or controllers to plan the driving route by using the touch screen. At the same time, the dual-lens camera can accurately identify road signs, and the car–car, car–road, and car–people interactions can be achieved with the help of the Internet of Things and mobile Internet to promote the free flow of traffic information sharing, ease traffic congestion, and improve the traffic network capacity.

12.2.2.2 *The development of the unmanned driving technology in China*

In China, the National University of Defense Technology (NUDT) developed the first smart car in 1989; in 1992, NUDT developed the first real unmanned driving car, and the research and development of the unmanned driving car technology in China officially started and entered the exploration stage.

The Cyber C3 smart car, an urban unmanned driving car project led by Shanghai Jiaotong University, mainly travels on unstructured roads such as non-main urban roads (road types and environmental backgrounds being complex, lane lines and road boundaries being unclear), with an average speed of less than 10 km/h, and it is capable of automatic positioning, navigation, driving, obstacle avoidance, and interaction.

Professor Zheng Nanning of Xi'an Jiaotong University set up the unmanned driving intelligent car project at the end of 2001, and they launched the unmanned driving car Siyuan No.1 in 2002, and in 2005, Siyuan No.1 successfully completed the campus road environment testing, and following that, the project team tried to get the Siyuan No.1 smart car out of the school. Then the development of the New Silk Road Challenge plan was made to get Siyuan No.1 to travel automatically from Xi'an to Dunhuang, but in the actual test, it was found that Siyuan No.1 found it difficult to adapt to the real traffic environment. After more than 10 years of research, Siyuan No.1 gradually accomplished a major breakthrough: it can travel from a specific section of road to a complex road, from a straight line to turn and to overtake, and the speed increases from 10–15 km per hour to 70–80 km per hour.

In 2011, the unmanned car FAW Hongqi HQ3 completed 286 km of road test; in 2015, the Baidu unmanned driving car completed the Beijing open highway autopilot test; and in 2016, the Changan unmanned driving car completed 2000 km of super driverless test, all marking the gradual maturity of China's unmanned driving technology.

Many other new technologies have emerged in the process of the rapid development of the unmanned driving technology. According to the Thomson Reuters Intellectual Property and Technology Report, 22,000 patents related to the unmanned driving technology appeared from 2010 to 2015, and many of these technologies have been put into use. Especially with the support of deep learning algorithms and cloud service technology, the development of the unmanned driving technology is progressing faster and faster.

12.2.3 *Global prospects for commercial applications of unmanned driving*

The initial manufacturing cost of the unmanned driving vehicles will be high because the unmanned vehicles have to be installed with radar, cameras, and artificial intelligence systems, and naturally the price will be very high, which ordinary consumers cannot afford. Therefore, the first field to use the unmanned vehicles will be some special industries and special groups, such as the courier industry, industrial enterprises, taxis, services for the elderly or disabled, and so on.

12.2.3.1 *Public transportation*

In the future, the public transportation industry will make use of the unmanned driving vehicles on a large scale. According to a study conducted by the University of Texas at Austin on the shared unmanned vehicle (SAV), one shared SAV can replace 11 conventional cars and increase operating mileage by at least 10%. This means that with the widespread use of the shared SAVs, traffic congestion can be effectively reduced, environmental degradation can be effectively controlled, and with convenience of use, the shared SAV will become the preferred mode of transportation for people.

In order to promote the unmanned driving vehicles, some cities plan to designate special blocks for the unmanned driving vehicles where the manned vehicles and the unmanned ones will not run at the same time, and the unmanned driving taxis and the shared SAV will assume the responsibility of providing all transportation services for the users. In order to create a good operating environment for the unmanned driving vehicles, the urban planning department will carry out area optimization.

12.2.3.2 *Express vehicles and industrial applications*

The unmanned vehicles will also be widely used in the courier industry among the queue truck field. With the rapid development of Internet e-commerce, the courier companies are rapidly emerging. The consumers shopping online always hope to get their goods in the shortest possible time, and especially for the fresh fruit and other food products, they hope that courier companies can provide the delivery service within a few hours to their homes. According to data provided by the Ministry of Commerce, in the first half year of 2018, China's e-commerce industry reached 4.08 trillion yuan in sales, whose total volume of express parcels exceeded the annual business volume of 2015, and many more e-commerce companies promised same-day delivery, further promoting the development of electric car and truck delivery.

The increase in the number of electric cars and trucks has increased the number of traffic accidents to some extent. According to a survey, in the US, the truck miles traveled account for 5.6% of the total motor vehicle miles traveled, but the truck-induced traffic fatalities account for 9.5% of traffic fatalities. In general, large trucks cost more than $150,000 to produce, and trucks can cost more if they are equipped with cameras and sensors. In comparison, the production cost of small cars is inherently lower. Therefore, due to the cost constraints, it will be difficult to achieve large-scale adoption of the unmanned driving trucks at the early stage.

12.2.3.3 *Elderly and disabled people*

At present, the unmanned driving vehicles have been used on a large scale for two groups of people: the elderly and the disabled people. Because of the physical limitations, the elderly and the disabled people face the problem of inconvenience, and the use of unmanned driving vehicles can help these two groups of people to solve their travel problems.

It is predicted that by 2050, the elderly population in the United States will exceed 80 million, accounting for more than 20% of the total population, and one-third of them will face travel problems, which is also the case in China. By 2050, China's elderly population is expected to account for 33% of the total population. In Japan, the proportion of elderly people in the total population is expected to reach 40% by 2060. At the same time, the size of the disabled population is enormous. For example, there are about 53 million people with disabilities in the United States, making up 22% of the adult population. About 13% of US adults have mobility impairments and 4.6% have visual impairments.

These people with disabilities and the elderly provide a wide market for the promotion of the use of the unmanned driving vehicles, and both groups seek independence and want to be able to travel freely. It is the unmanned vehicles that can precisely meet their needs, allowing them to live independently and maintain an active and optimistic attitude toward life.

12.2.4 *World's leading companies' unmanned driving car research and development*

In recent years, the development of the Internet, artificial intelligence, and other technologies has laid a technological foundation for the development of the unmanned driving cars, and the world's leading Internet companies and automotive giants have built certain competitive barriers to the development of the unmanned driving cars by virtue of the technological advantages they have accumulated in the unmanned driving field.

12.2.4.1 *Tesla planning to realize full unmanned driving*

The electric car manufacturer Tesla has a leading position in the field of driver assistance technology and unmanned driving technology, and its CEO Elon Musk believes that there is wide scope for complete unmanned driving technology, that fully unmanned driving can be achieved in 2 to 3 years, while regulatory approval is still 1 to 5 years away. In October 2015, Tesla installed Autopilot software on the Model S, which allows the vehicle to steer, change lanes, and stop automatically, but Tesla was criticized in the summer of 2016 for the first traffic fatality in Autopilot, although no safety flaws were found in Autopilot itself.

After the accident, Tesla severed its relationship with Mobileye and has begun working to increase its control over the development of radar and camera systems. Tesla has promised the public the semi- and fully unmanned driving services. It has received 400,000 orders for its Model 3. Tesla has also declined to cooperate with Uber to develop unmanned driving vehicles.

Since October 2016, all cars produced by Tesla have been equipped with a new sensor and computer software package, Autopilot Hardware 2, which will allow the cars to be fully driverless when the software are mature. The system replaces Mobileye's EyeQ3 with Nvidia's DrivePX2, and the users will have to pay to activate them in order to enjoy fully unmanned driving function. According to users' feedback, the initial application of the Autopilot 2.0 software was ineffective, but the system has since been updated and improved.

Tesla continued its previous style of doing things by promising the public an aggressive timeline for delivering the unmanned driving functionality and releasing the feature by the end of 2017. The company reversed its previous data collection strategy and used customers' in-car cameras to collect video information to create the new system. The Tesla Unmanned Driving User Agreement also makes the following stipulation that the customers who want to benefit from the sharing vehicles can only trade on the Tesla network. In addition, Tesla has purchased SolarCity to create a complete sustainable transportation ecosystem.

12.2.4.2 *Volvo planning to achieve automated driving by 2021*

In the field of automated passenger cars, Volvo has made great progress. Volvo has enjoyed an excellent reputation in the field of safety technology innovation and calls the technologies related to automated vehicles the "smart safety" in the hope of using these features to reduce the incidence of fatal accidents in Volvo vehicles to zero. In the past, in 2017, Volvo successfully provided 100 customers in Sweden with an XC90 SUV with autopilot functionality, though with a number of restrictions in terms of autopilot zones, times, and conditions.

Volvo promised that it will take full responsibility if the vehicle is involved in an accident while on autopilot, and announced the extension of the autopilot program to the United States and China. Volvo is keeping pace with its competitors such as BWM and has a deadline of 2021 to achieve full deployment of automated driving vehicles.

While keeping up with its competitors, Volvo is also actively looking for partners to partner with Sweden's Autoliv to form an automated driving joint venture, Zenuity, which plans to be the first to commercialize driver assistance systems and make them available to other automakers in 2019.

12.2.4.3 *Baidu planning to spin off its automated driving car business for mass production*

From 2015, Baidu conducted public testing of the unmanned driving technology. In November 2016, Baidu invited the public to ride in an electric, homemade unmanned driving car to experience the testing of unmanned driving technology, which lasted for a week. At the same time, Baidu also obtained a California driverless car autonomous testing permit. In June 2016, Baidu announced that it wanted to enable mass production of automated driving cars within five years, for which it set up an AI research lab in Silicon Valley. In September 2018, Baidu launched a fund project for the development of unmanned driving mobile, with a total project value of $1.5 billion. Baidu headquarters plans to produce a limited number of unmanned driving cars, and plans to mass-produce unmanned cars in 2021.

12.2.4.4 *Audi creating automated driving subsidiary*

Audi has released a number of automated driving car prototypes on the basis of the A7 and RS7 models, and some of the vehicles have already begun testing for consumers. At the end of July 2016, Audi created a subsidiary of its own, SDS, specifically to develop the unmanned driving technology, through which Audi joined the competition in the unmanned driving field. In April 2017, Audi hired the former head of Tesla's auto-navigation program as CTO of the Autopilot division, underscoring Audi's commitment to enter the unmanned driving field.

According to Audi's development plan, the automatic driving technology in the A8 flagship sedan will be commercialized. The automated driving function at Level 3 is defined as "autopilot under certain conditions", where commercial application needs to be approved by the regulatory authorities. As the luxury brand of Volkswagen, Audi's entry into the autopilot field is significant.

Audi, Daimler, and BWM have cooperated to form a consortium which spent $3.1 billion to buy Nokia's HERE Maps precise positioning assets. In recent years, HERE is accelerating its plans to standardize the data collection and transmission of vehicle sensors.

12.2.4.5 *Microsoft seeking cooperation with automakers*

Microsoft has also been involved in automated driving car research, and at the initial stages, Microsoft's strategy was to collaborate. For example, in November 2015, Microsoft struck a deal to use its HoloLens technology to work with Volvo on the development of automated driving vehicles.

In March 2016, Microsoft announced an expansion of its five-year partnership with Toyota to provide strong support for Toyota's research on robotics, automated driving, AI, and other areas. By June 2016, Microsoft's strategy remained to provide technical support to partner companies and it has not yet launched its own research and development strategy. Microsoft's own Azure cloud business has seen rapid growth driven by its automotive customer business. According to related reports, Microsoft has also been involved in the previous HERE HD map service mastered by BMW, Daimler, and Volkswagen and holds certain shares.

12.2.4.6 *Toyota's "guardian angel"*

In 2014, Toyota issued a statement claiming that it would not develop the unmanned driving cars for safety reasons, which has since been reversed, and in 2015 Toyota announced that it was investing $1 billion in the development of unmanned driving cars and had established the Toyota Research

Institute (TRI) for the study of the unmanned driving technology, employing professors and researchers from Stanford University and MIT and inviting the unmanned driving car company Jay bridge Robotics to join them. In April 2016, Toyota announced a partnership with the University of Michigan to co-develop unmanned driving cars, which was the third university cooperating with Toyota.

Toyota planned to distribute research and development tasks of the unmanned driving car among the three universities, with the University of Michigan being responsible for the fully unmanned driving car research and development, Stanford University for partial unmanned driving car research and development, and MIT for research on machine learning algorithms. In August 2016, Toyota increased its investment in each university, investing $22 million in the University of Michigan in order to promote robotics and research on the unmanned driving technology.

Toyota plans to get the AI car to work on road in 2021. TRI CEO Gill Pratt has been a big supporter for the "Guardian Angel" system, which only works when a human driver is about to be in danger or makes a wrong operation and it will monitor and intervene in the driver's behavior. Currently, TPI is working on a second system, Chauffeur, aiming at the unmanned driving at Level 4 and 5 and the system will be deployed for applications after Guardian. In March 2017, the institute presented an updated autonomous development platform to the public.

12.3 Technology Routes: Key Technologies for the Unmanned Driving Industry

12.3.1 *Key components of the unmanned driving technology*

In recent years, universities, research institutions, and technology companies have actively explored the field of the unmanned driving, creating many unmanned driving automobile projects, and the unmanned driving car market has entered the explosive growth stage. According to the *Unmanned Driving Car Industry Development Prospects and Investment Strategic Planning Analysis Report* published by the Prospective Industry Research Institute, the global unmanned driving car market size was $4 billion in 2016; in 2018, the global unmanned driving car market size was $4.82 billion; and by 2021, the global unmanned driving car market size will reach $7.03 billion. In order to promote the rapid development of the unmanned driving industry, all countries are focusing on the research on other related technologies.

The unmanned driving technology is a large and complex system which realized the organic integration of multiple fields and disciplines, including the following.

12.3.1.1 *The systematic structure of automotive*

The systematic structure of automotive builds the skeleton for the automotive system and defines the principles of the unmanned vehicle system's hardware and software organization, integration methods, and support procedures.

12.3.1.2 *Perception and identification of external environment*

The artificial intelligence system is a data-driven system, and the unmanned vehicle system is no exception. The core means for the unmanned vehicles to perceive the environment and grasp real-time road condition information is on the basis of the external environment perception and recognition technology of on-board sensors. With the support of the external environment perception and recognition technology, the unmanned driving car can let the on-board intelligent system obtain the relative position, distance, speed, and other information between the objects on the road (for example, pedestrians, cars, obstacles, etc.) in real time to provide the necessary support for the car's decision-making and command execution.

12.3.1.3 *Position navigation system*

During the driving process of the unmanned driving cars, the positioning and navigation system will provide a variety of information such as car position, speed, direction, etc., which is the basic software for the unmanned driving cars. After years of development, the positioning and navigation technology system has been constantly improving, and a variety of positioning and navigation technologies such as satellite navigation technology, roadmap positioning technology, visual positioning technology, track estimation technology, map matching technology, inertial navigation technology, and so on have emerged. As a matter of fact, two or more than two positioning and navigating technologies are applied at the same time in the real situation so that the accuracy and efficiency of navigation will be improved.

12.3.1.4 *Path planning technology*

The path planning technology in the unmanned driving field refers to the technology by which the unmanned driving car's intelligent system will design a safe and efficient driving route between the established starting point and destination, by analyzing data such as road conditions, speed, distance traveled, number of signal lights, etc., which requires the application of statistics, big data, cloud computing, and other technologies.

In terms of the analysis of path, the driver can make flexible decisions in combination with real-time road conditions while driving a car, but in an unmanned driving state, the intelligent control system needs to make use of a high-precision 3D road condition model to efficiently calculate the relative position, distance, speed, and other information about the car and surrounding things, so as to achieve safe driving.

The road condition thinking model needs to be adjusted in real time, and this process will involve massive amounts of data, which must be processed efficiently with the help of GPU or distributed neural network algorithms. It should be pointed out that in order to update the high-precision three-dimensional road map in real time, it is necessary to share data between the intelligent systems of automated vehicles in the region. In order to obtain richer and more diverse information about road conditions, the automated vehicles need to install software and hardware devices such as laser distance radar, inertial radar, and various sensors.

In terms of path planning, the path planning not only requires high-precision three-dimensional road maps, but also has a high dependence on high-precision positioning technology. The unmanned driving car must be able to perceive and identify its own position in real time, execute the optimal route scheme set by the intelligent system, and comply with the traffic order, avoid obstacles in time, and adhere to the safety of people as the first priority to complete the driving task safely and efficiently.

The route planning function mainly consists of three points:

(1) The carriageway is planned by taking the overall environment into account to design the optimal route between the departure point and the destination and taking the individual needs of passengers into account as well.

(2) The optimal driving lanes are selected by the integration of lane information and by using navigation, sensors, intelligent algorithms, and other means and tools.
(3) The conditions formed by understanding the lane information, making full use of the deep learning algorithm, analyzing the specific requirements of the lane on the speed, steering, and other aspects allow the intelligent system to make more scientific and reasonable decisions when controlling the vehicle driving.

12.3.1.5 *Vehicle control technology*

On the premise of acquiring environmental information, the vehicle control technology should be used to keep the car in a smooth exercise state by using the automatic steering control system; and in the specific exercise of the road, the vehicle control technology can not only control the car's travel speed and distance, but also allow the car to complete overtaking, lane change, and other driving operations. In order to achieve these diverse functions, it is necessary to install control and operating systems in the unmanned driving cars, such as satellite navigation systems, adaptive cruise control system, emergency braking system, lane departure system, automatic parking system, and so on.

(1) Emergency braking and satellite navigation system: When there is an emergency situation, the braking system will be activated immediately in response to ensure that the unmanned driving car is under the effective monitoring of the human.
(2) Adaptive cruise control system: This system is to ensure that the unmanned driving car maintains a safe distance from the vehicle in front of it so that the car can be safely exercised on the road.
(3) Lane deviation detection system: This system is to ensure that the car can follow the road signs and sound guidance to exercise. When signs of deviation from the lane are found by the vehicle, it will make rapid direction adjustments to avoid the wrong route of the car.
(4) Automatic parking system: This is the system to ensure that the unmanned driving car can automatically stop at the correct location and can be safely driven out.

These technologies interacting with each other will ensure that the unmanned driving vehicles can work safely with purpose and direction.

The above technologies may seem isolated from each other, as a matter of fact, they can be systematically grouped into three levels: the levels of perception, decision-making, and execution. Among them, the perception and identification of the external environment is classified into the perception level, the positioning and navigation system and path planning are classified into the decision-making level, and vehicle control technology is classified into the executive level. In addition to these main technologies, there are many other technologies involved in unmanned driving, which are detailed as follows.

12.3.2 *Perceptual level: Collecting environmental and driving information*

The unmanned driving cars make full use of the collaboration among radar, sensors, satellite navigation systems, artificial intelligence computing modules, and other hardware and software facilities to allow the car to drive autonomously without anyone interfering. The unmanned driving car will be equipped with a large number of on-board sensors, which are equivalent to equipping the car with a myriad of eyes, therefore, the unmanned driving car is significantly safer and more reliable than ordinary cars.

In the process of the unmanned driving, the most important thing is the implementation of the unmanned operation on the basis of the perception of the environment and this technology plays a fundamental support role in the unmanned driving technology. In addition to conducting environmental perception, the application of vehicle control technology is also involved in the technology, that is, the development of the unmanned driving has to depend on environmental perception and vehicle control technology support.

In a relatively well-developed unmanned driving system, the information collection network collects information about the car's surroundings, and then an intelligent algorithm formulates a driving plan, followed by the use of an intelligent vehicle driving controller to get the car to its destination.

The perception level is an important channel for the unmanned driving cars to obtain external information. At present, the intelligent sensor system for the unmanned driving cars to perceive the external environment mainly contains three components: visual sensors, LIDAR, and millimeter wave radar. The navigation through the visual sensor is also known as visual navigation. The vehicle will achieve car positioning and detecting the surrounding environment by using the camera to obtain local images of the environment around the road and using the technology of image recognition, etc.

The lidar allows the unmanned driving cars to monitor dynamic obstacles in real time and it is characterized with anti-light interference ability and high detection accuracy, which can provide strong support for dynamic obstacle monitoring and motion status assessment. The lidar is able to acquire many environmental parameters such as the number of lines, accuracy, point density, scanning frequency, detection distance, horizontal and vertical viewing angle, providing basic information such as position and distance, as well as feedback on the density of the target object, thus allowing the algorithm to derive its reflectivity, which is helpful for in-depth analysis.

The millimeter-wave radar has a frequency range between 30 to 300 GHz, which is similar to the low frequency band or the centimeter band, and it is able to make all-weather images. The high-frequency band takes full advantage of the high resolution of infrared waves. The millimeter-wave radar and lidar follow the same detection principle, and they use the reflection of the object on the radar wave to obtain the target object data, but the millimeter-wave radar technology is more mature. In terms of penetration and cost it has certain advantages, but the disadvantage is also more prominent, for example, the frequency band on the detection distance is too large, it is difficult to perceive the pedestrians and it cannot realize the target of fine identification.

The mainstream trend of the unmanned driving car exploration projects is to improve the car's ability to sense its own operating state and the surrounding environment through the integrated use of a variety of intelligent sensing systems. In general, the process of unmanned driving consists of two steps: the first step is to obtain the information about the environment by using technologies such as radar; and the second step is to control the car's braking and steering systems on the basis of the judgments about the environment to ensure that the car can safely and steadily function on the road. By combining these two modules and taking advantage of the synergy between them, the entire process of the unmanned driving can be accomplished.

The car can capture information about its current location, the road, surrounding vehicles and obstacles, and send this information to the central computer in the car by using the environmental sensing technology so that it can grasp the situation on the way to help the car to set and adjust the driving route, and finally reach the planned goal. In other words, the car can obtain information about its location and surrounding objects through the environment perception technology, judge the properties of these objects and their distance from itself, and grasp the current environmental information from the macro level.

At the present stage, the technologies used in the environmental sensing mainly include visual sensing, laser sensing, and microwave sensing. With the help of these technologies, the car sensors, in-vehicle communication systems, and radar systems can capture the environmental information, present two- or three-dimensional images by using information technology, and send the acquired information to the information processing terminal, where the technicians will process the three-dimensional images to perceive and judge the real-time environment of the car. In addition, different cars can also be connected to collect real-time environmental and driving information of different vehicles. The application of remote communication technology and Li Yongle's wireless network technology in the unmanned driving field effectively improves the car's ability to perceive the environment and promotes the development of unmanned driving as a whole.

12.3.3 *Decision-making level: Route planning and real-time navigation*

The main function of the decision-making level of the unmanned driving vehicle is to plan the driving routes and provide real-time navigation services. The realization of these two functions requires the use of high-precision maps, also known as "high-definition digital maps". The difference between high-precision maps and ordinary maps is that the former is more accurate and informative, and it is used exclusively for the unmanned driving cars.

The reason why the unmanned driving cars cannot make use of ordinary maps is that the ordinary maps are marked with simple information, with two parallel lines representing roads and intersections representing crossroads. People can use their cognitive abilities to make judgments on the basis of this simple information, but it is difficult for the unmanned driving cars to function in this way. Therefore, the high-definition digital maps are three-dimensional and cover a wealth of information, including lane lines and coordinate locations of nearby facilities, and they are extremely accurate to the centimeter level.

Compared to the ordinary electronic maps, another special feature of the high-precision map is its ability to collect the intensity of reflections from road lidar. For the human drivers, this feature is meaningless, but for the automated vehicles, it is extremely valuable. The intensity of road lidar reflections is a very slowly changing and small road feature that can provide a basis for positioning an unmanned vehicle by using optical radar. By comparing the information obtained using the optical radar scans with the information already available on a high-precision map, the unmanned driving system will be able to pinpoint the current location of the vehicle.

Two conditions are needed for the unmanned driving decision-making system to work: a stand-alone intelligent vehicle and an intelligent transportation system, such as V2V, and V2X is another technology that allows for the path planning in addition to high-precision maps, and it is based on V2I. V2X allows vehicles to connect to their surroundings, forming an Internet of Things that covers vehicle-to-vehicle communication systems, vehicle-to a range of communication systems, such as infrastructure communication systems and vehicle-to-pedestrian communication systems.

Running a red light can be avoided completely if the vehicle can "get" the signal light information while driving, rather than just "see" the signal light information. Here the term "get" means that when the sensor has not yet "seen" the signal light, such as in the distance of 100 meters from the signal light, the signal light actively sends to the vehicle the current signal status, changes duration, and other information to remind the vehicle to slow down, accelerate, or brake and thus to avoid running red lights. In addition, if the unmanned driving vehicles can "get" the driving intentions of surrounding vehicles while driving, such as stopping, changing lanes, turning, etc., they will effectively avoid traffic accidents.

With the support of high-precision digital maps and the V2X communication network, the automated driving system can use search algorithms to accurately assess the cost of various driving behaviors and thus to plan the optimal driving path.

12.3.4 *Intelligent transport executive level: Precise control of the car's operation*

In the entire control system, the executive level is located at the bottom, and its main function is to perform the operations such as braking, steering, acceleration, etc. The system uses the "wire control device" to control the steering wheel and throttle, and configures the subsystem composed of multiple processors to carry out stable and accurate control of the entire mechanical system of the automatic driving vehicle.

The car driving controller is the core module that realizes vehicle control and automatic driving, and it includes two kinds of functions: on the one hand, it will complete the driving task with safety, efficiency, and low energy consumption through the combination of the dynamics of the automatic driving car and the performance of the sensor in the car to adjust the vehicle driving parameters; on the other hand, the car will be able to think and make decisions like a driver with experience of many years on the basis of the dynamics of the unmanned driving car and the sensor recognition performance, and the car may also take the safety, economy, comfort, ethics, and many other factors into account and adjust the driving state in accordance with the real-time road conditions, which, at the same time, will bring passengers a good travel experience and maintain the harmony between the vehicle and other vehicles and pedestrians.

The precise control over the vehicle is the key to automated driving, and the basic principle of the vehicle control is to ensure the safety of people and vehicles, which obviously requires the strong support of advanced braking systems. When a car is involved in an accident, the braking system is able to apply precise, real-time braking at the behest of an intelligent driving controller to avoid or reduce the negative impact of the accident. With a strong braking system in place, an intelligent vehicle also requires more intelligent control algorithms, which are the primary means of giving the braking system greater control, precision, and robustness.

At this stage, almost all the self-driving car projects may make use of technologies such as machine learning to allow the car to learn and simulate human driving experience, which is then combined with intelligent systems to analyze and control the vehicle in real time. However, if the unmanned driving cars have the ability to deal with emergencies, they will effectively control risks, and the intelligent systems for dealing with emergencies will not work like people who may generally feel nervous and have panicked emotions, as the unmanned driving car is rational and decisive in decision-making, of course, the premise of which is the strong "intelligent brain".

Safety is the primary consideration for an unmanned driving car. In the state of artificial driving, there are certain differences among the driver's driving skills, psychological quality, emergency handling capacity, operating habits, etc. When the driver encounters emergencies, s/he can take differentiated response measures, but the unmanned driving cars are emotionless and the guarantee of the safety should be the first priority. Then we should take the experience of using the car into consideration. At present, the smart cars in the market are mainly semi-automated cars that provide driver assistance services, so ensuring the passengers' comfortable experience of the unmanned driving cars is an important issue in the future.

In the future, the mature artificial intelligence will be able to analyze the surrounding environment and make decisions like a person. When the driving strategy in a particular scenario is analyzed, it does not have to re-analyze every time, and it can make use of the car's previous strategy for handling the scene to control the vehicle, or imitate the vehicle in front, which will have a disruptive impact on the automotive industry, and human life will also change significantly.

Theoretically, the unmanned driving cars will be equipped with the conditions to "drive" through the combination of the sensing, decision-making, and execution systems. As the authors of *Driverless: Intelligent Cars and the Road Ahead*, Hod Lipson and Melba Kurman, put it, "Although the technology is almost ready, the social context in which this unique technology will be deployed may not be. For example, the laws and regulations governing unmanned driving are not yet in place. However, the unmanned driving cars are bound to increase in number and the era of the unmanned driving cars will eventually come on the basis of such advantages of unmanned driving cars as their efficiency and safety. We believe that the development trend of 'unmanned vehicles' is irresistible".

12.3.5 *Application of unmanned driving in the field of intelligent transportation*

The unmanned driving technology includes the human–computer interaction technology, intelligent control systems, anti-lock braking systems, traction or stability control systems, etc. A detailed analysis of the application of these technologies in the field of intelligent transportation is conducted as follows:

(1) Human–computer Interaction Technology Ensuring Driving Safety: The intelligent transportation systems takes human safety as the first priority, and the car and the human need to interact with each other to

ensure the human safety in all aspects, which requires the use of human–computer interaction technology. The human–computer interaction involves not only the interaction between the car and the human inside the car, but also the interaction between the car and the humans outside the car. At this stage, the research on human interaction technology in the field of the unmanned driving cars focuses more on the interaction between the car and external pedestrians, and the efficient collaboration between the car and the road can be achieved by building a communication mechanism that allows the car and people to interact efficiently and conveniently. In the long run, the unmanned driving cars in the future will mimic human behavior patterns when making intelligent decisions.

The efficient and convenient human–computer interaction can bring users a better travel experience, which has a very positive impact on the promotion of the market ability of the unmanned driving cars. In the absence of human–computer interaction, it is difficult for the cars to fully understand the needs of people. The unmanned driving cars will analyze massive amounts of data in real time for the route planning, lane changing, obstacle avoidance, parking, and so on.

(2) Intelligent Control Systems Improving Travel Experience: The intelligent decision-making from the vehicle is an important assessment index of the performance of unmanned driving cars. The data obtained through sensors and other devices will be processed by the processor, and the application of data models and intelligent algorithms to achieve optimal decision-making. Of course, this intelligent algorithm should have self-learning and self-improving capabilities so that it can meet the actual needs of the unmanned driving cars in a variety of driving scenarios. The automated control is the executor of decision-making instructions, allowing the car to make a series of behavioral actions, for example, acceleration and deceleration, turning, obstacle avoidance, etc., which has a direct impact on the human travel experience.

In many traffic accidents, the traffic accidents caused by the human factors occupy a fairly huge proportion because people will feel tired and have psychological fluctuations if they are stimulated by sound and images or other external factors, which thus will result in human errors, threatening the safety of people and property. The intelligent control system based on the unmanned driving technology can maintain the objectivity and calmness for a long period of time without any sense of fatigue, which can effectively reduce the number of man-made traffic accidents.

(3) Anti-lock Braking System Reducing Traffic Accidents: The anti-lock braking system (ABS) is one of the relatively mature unmanned driving systems that is widely used. The tires of a vehicle which is not equipped with ABS may be locked up after an emergency braking accident, which will cause the car to skid, and an experienced driver may avoid locking up the tires by stepping on the brakes repeatedly, however, the majority of the drivers are easily distracted in a few seconds and even treat the accelerator as the brake. When the car is equipped with the anti-lock braking system, in the event of an accident, the system can take effective measures to reduce the possibility of traffic accidents as the system monitors the tire status in real time before the tire is about to lock up.

(4) Traction or Stability Control Systems Supporting the Autonomous Movement of the Vehicle: The traction or stability control system is also the technology which is widely used in the unmanned driving system. Compared with the anti-lock braking system, this system is more complex and it can monitor information such as road conditions, vehicle speed, driving direction, and operating status of key components in real time so that effective measures can be taken in time to ensure the driving safety when it is found that the car may lose control and roll over. In the true sense of unmanned driving, the car can efficiently and safely complete the tasks given by passengers or controllers independently.

(5) The Interaction of Each System Truly Realizing Unmanned Driving: The traffic information collecting system, information processing and analyzing system, and information disseminating system are the intelligent transportation systems which are closely related to the unmanned driving technology. Among these systems, the traffic information collecting system is built on the basis of the global positioning system (GPS) technology as well as a variety of systems including infrared radar detection system; the information processing and analyzing system is usually applied to the GIS application system, expert system, and other systems; and the information release system depends greatly on the mobile Internet, mobile phone terminals, and other media technology.

The developed countries like the United States, Japan, and other countries have taken the lead in creating unmanned systems as the core of the intelligent transportation system by virtue of their leading edge in computer control, network technology, positioning and navigation, artificial intelligence, and other fields, with the result that the unmanned driving car projects have emerged in large numbers, and it will cost them less in the exploration and practice.

12.3.6 *The impact of unmanned vehicles on urban traffic flow*

In recent years, with the increasing number of the urban population and private cars, the problem of urban traffic congestion is becoming more and more serious. In order to plan urban road networks better and make the scale and form of urban road construction more reasonable, people began to analyze the efficiency of urban traffic and its influencing factors. There are various factors which may affect the traffic flow, and what role do the unmanned vehicles play in these factors?

The unmanned vehicle is equipped with on-board sensors to collect environmental data around the vehicle. The on-board sensors are mainly composed of cameras, global positioning systems, lidar, and other parts. At the same time, the intelligent processing module in the unmanned vehicle can effectively control the operating parameters of the unmanned vehicle, such as steering parameters, speed parameters, and so on, to ensure that the unmanned vehicle can drive safely.

Specifically, the intelligent processing module is composed of the decision-making module, control module, and other parts, which can obtain a variety of data, including video image data collected by the camera, point cloud data collected by lidar, position data collected by GPS, steering data or speed data output by the decision-making module, execution result data output by the control module, etc. The intelligent processing module is used to control the operation of unmanned vehicles.

In many areas, the traffic capacity is very limited due to the number of road strips and other reasons. Unmanned vehicles achieve automatic driving, which can reduce the distance between vehicles, ensure the safety of vehicle driving, reduce the occurrence of traffic accidents, increase the capacity of road traffic, and improve the road capacity.

12.3.6.1 *Analysis of factors related*

There are many factors that affect the capacity of motorways, from which three factors are selected for the analysis, namely the number of routes, the proportion of the unmanned driving vehicles among all moving vehicles, and the peak capacity of the road to obtain their correspondence.

The unmanned driving vehicles do not have a driver, and the car's reaction time is close to zero if there is an emergency on the road. According to the data, the average speed of the unmanned vehicles is much faster than the average speed of the manual drivers, so the higher the percentage of the unmanned vehicles among all the vehicles in motion, the higher the road's capacity will be. Also, the greater the number of road routes, the greater the capacity of the road will be. In addition to these two major factors, whether the road has intersections also has a significant impact on the road's capacity to pass, and the more the number of intersections, the worse the road's ability to pass will be.

The traffic flow is not only affected by the road capacity but also by other factors, such as the geographic location, road network density, road service level, regional vehicle ownership, passing time, etc. These five factors will be analyzed in detail. The more prosperous the area where the road is located, the greater the road network density, the higher the road service level, the more the regional vehicle ownership, and the greater the traffic flow during the peak travel period will be.

12.3.6.2 *Analysis of the introduction of the unmanned driving vehicles on different roads*

Areas with different road network densities require different approaches to the introduction of the unmanned driving vehicles. First of all, the road capacity is affected by a variety of factors and if the situation of the roads is not fundamentally changed, the passing capacity of the roads will be significantly affected as more and more unmanned driving vehicles take to he roads. In order to gain insight into this impact, studies have been done by organizations and the findings are as follows.

Firstly, on smooth roads where the proportion of the unmanned vehicles is less than 54%, the growth rate of road capacity will continue to slow down as the proportion of unmanned vehicles keeps increasing; and where the proportion of the unmanned vehicles is greater than 54%, the growth rate of road capacity will accelerate as the proportion of the unmanned driving vehicles keeps increasing, but considering the manufacturing price of the unmanned driving vehicles, the unmanned driving vehicles are not suitable to be placed on such roads.

Secondly, in the case of a generally smooth road, where the proportion of unmanned vehicles is 40%, the unmanned driving vehicles will have the greatest impact on road capacity, and this impact will become weaker and weaker as the proportion of the unmanned driving vehicles continues to increase. Therefore, the unmanned driving vehicles can be introduced on such roads to relieve the traffic pressure.

Lastly, on the roads that are highly prone to congestion, the roadway capacity is strongest when the proportion of the unmanned driving vehicles reaches 45%. As the percentage increases, the passing capacity of the road will decrease. The reason for this phenomenon is that the maximum capacity of the road prevents the traffic from growing, and for these types of road sections, the traffic authorities can dedicate the unmanned driving lanes to alleviate the traffic pressure.

Comparing the congested road with the actual map will show how the traffic flow in the town center is the largest, especially at the intersection of the main roads with dense surrounding roads, therefore, the problem of traffic congestion can be solved if the unmanned driving lane can be opened in this area.

12.4 Intelligent Traffic Management Model Based on the Unmanned Driving

12.4.1 *Deep integration of free travel and shared modes*

If unmanned driving cars must replace manned driving cars, we need not only to break the barriers among Internet of Things, big data, cloud computing, sensors, artificial intelligence, and other technologies, but also make breakthroughs in urban planning adjustments, the improvement of traffic rules, and the construction of regulatory systems and other management aspects. From the development practice of many industries, technology and management complement each other, and only by the collaborative development of the two can the unmanned driving cars have large-scale promotion of popularization possible.

Therefore, in the exploration of traffic management models matching with unmanned driving cars, it is too late to explore them when the unmanned driving technology is fully mature, and the backward planning, construction, supervision, and other aspects will bring many restrictions to the development of technology. For example, setting a high threshold for the unmanned driving cars on the road will result in its inability to conduct road tests and bring major obstacles to the development of unmanned driving technology.

Many countries around the world are now aware of the important value of intelligent traffic management models. For example, in August 2014, the Singapore government established the Singapore Automated Road Traffic Committee to provide guidance and assistance to the development of the unmanned driving vehicles in four major dimensions: freight, fixed routes, point-to-point transport, and public facility operations.

(1) Freight: The transportation and delivery of goods is accomplished through an automated fleet of trucks.
(2) Fixed Route: Transportation services are provided on public transportation routes within and between cities.
(3) Point-to-point Transportation: The demand for point-to-point and short-range trips is met through the sharing locomotive.
(4) Operation of Public Facility: The public services such as plant irrigation, garbage collection, and road cleaning are provided.

In December 2017, two guiding documents, namely, *the Guidance Opinions on Accelerating the Work Related to the Road Test of Automated Vehicles* and *the Implementation Rules for the Management of the Road Test of Automated Vehicles,* were issued by the Beijing Municipal Transportation Commission, in conjunction with the Municipal Public Security and Transportation Bureau and the Municipal Economic Information Commission, which will undoubtedly have a very positive impact on the further development of the unmanned driving vehicle technology and management models.

The establishment of intelligent traffic management model should be based on people's travel patterns, and needs to highlight its humane and intelligent considerations. From the perspective of the number of travel users, people's travel patterns can be divided into two categories.

(1) Free Travel: In the mode of free travel, the individual travelers have completely independent control and use of travel tools and they are free to make allocations of its internal space facilities, which can meet people's individual travel needs.

(2) Shared Travel: In the mode of shared travel, two or more travelers use a travel tool together and they influence each other. The distribution of internal space and facilities is subject to certain constraints, and it also involves certain privacy protection issues. Strictly speaking, only individual self-driving trips are considered free trips.

There is a special situation common to free travel and shared travel. When a person takes a taxi alone, the taxi driver is a service provider just like the taxi, who is not a shared user. The taxi driver himself is an independent person who has independent awareness, and the passenger cannot ignore his existence, so his freedom will be restricted. The safety of passengers and property may be compromised if they encounter an unscrupulous taxi driver.

The advantages of free travel are mainly reflected in the convenience, safety, and comfort, while the advantages of shared travel are mainly low-cost and green or environment-friendliness, which can effectively reduce the total number of cars in the city, ease traffic congestion and parking problems, and improve efficiency of resource utilization. Under certain economic conditions, most people tend to choose free travel, and the awareness of sharing needs to be improved.

From the perspective of the ownership of travel tools, we can further divide people's travel mode into two categories: private car travel and public travel. When the ownership of the vehicle used by people does not belong to the users, it is considered a means of public travel, for example, taxi, online car, bus, subway, commuter bus, station wagon, and using the private car of family members or friends.

Public trips in the manned driving state and the trips with two or more people sharing a private car are not considered free trips. After the mature development of the unmanned driving technology, riding in unmanned driving taxis will be similar to the travel by car and it will become truly free travel.

In other words, the unmanned driving taxis will realize the deep integration of free travel and public travel, allowing people to enjoy the advantages of convenience, comfort, and safety of free travel, and gaining the low-cost, green advantages of public travel as well. Considering the relatively high price and limited users of unmanned driving cars in the early stage of development and the fact that the unmanned driving taxis can reduce labor costs, it becomes a natural thing for the unmanned driving taxi to be the first to be used on a large scale in the market.

12.4.2 *The unmanned driving taxi and the free travel model*

In August 2016, the taxi startup nuTonomy in Singapore conducted a trial run of the automated taxis in Singapore, which allowed eligible passengers to access the service via their smartphones. Six unmanned driving taxis were put into service during the trial phase, with 12 passengers qualifying for the experience, and the taxi travel was restricted to a 2.5 square mile commercial and residential area and the passengers were required to be in certain locations to get on and get off. As per nuTonomy's plan, the total number of cars in Singapore will be reduced from 900,000 to 300,000 in the future as more and more unmanned driving cars are put into use. In October 2017, Delphi Pike Electric announced that it had completed the full acquisition of nuTonomy for $450 million.

For a country as small and economically developed as Singapore, the difficulty in the promotion of the unmanned taxis is relatively small, and their application will bring many conveniences to people's travel generation, which will have a very positive impact on the improvement of traffic congestion. It should be pointed out that public transport is the core component of Singapore's transport system, and the application of the unmanned driving taxis is not to completely replace the traditional taxis, but to further improve it. In the future transport development plan released by Singapore, it is expected that by 2025, public transport will be responsible for up to 75% of peak travel services in Singapore.

The companies such as Uber have been testing unmanned driving cars not only in Singapore, but in various cities such as California, Michigan, and Pittsburgh to try and launch the unmanned driving taxi services. Especially for the network car company such as Uber, after the application of the unmanned driving taxi, the platform is able to customize and design a perfect travel plan for people's daily travel. People's travel demand can be effectively predicted through the intelligent systems for efficient allocation of vehicle resources, and more importantly, through the combination with the big data and cloud computing technology. The targeted adjustments to the allocation of transport resources in different areas can be made so as to better meet the peak travel demand and to reduce the idling rates.

The slogan of the algorithm competition organized by Didi Research Institute in cooperation with Uda Xue Cheng and Zhihu goes,

"The dream of changing the world can come true, your algorithm will be used in the real travel world, and maybe one day you can say, 'In this city, my algorithm is in charge of your choice of which car to take'". If the unmanned driving taxis can be promoted and popularized on a large scale, people will no longer need to buy, repair, and regularly maintain private cars, which will eliminate all kinds of problems such as parking difficulties, car insurance, traffic jams, breakdowns, and accidents on the way, which is highly in line with people's aspirations for a better life.

Children, the elderly, the people with disabilities, and those who do not have a driving license can also enjoy the freedom of travel through the unmanned driving taxis, which will greatly reduce the number of private cars and help protect the environment, reduce traffic accidents, and improve travel efficiency and experience.

Scientific research has found that if the unmanned driving taxis can be promoted as the main mode of urban motorized transport, and the number of private cars is strictly limited, then the total number of vehicles in the city will be reduced by nearly 50%, with 40% less parking space occupied, and the accident rate would be reduced by about 90%. If the unmanned driving cars can be upgraded to be electric and other new energy vehicles, the vehicle exhaust emissions will be reduced by about 80%. In this case, the city will have more space to improve the quality of life of residents, for example, the building of more parks, soccer fields, and other public recreational facilities; and the bicycle lanes can be widened as well.

In order to solve the problem of air pollution in the city center, Oslo, in Norway, will achieve "Zero Private Car" before 2019; Madrid, in Spain, announced that by 2020 the vehicles will be banned from entering the city center for about 2 square kilometers. Of course, in order to meet people's individual travel needs, the private cars will not disappear for quite some time, but the freedom of experiencing driving a private car does not necessarily mean having to buy a private car, and in the future, there may be some providers of long-term rental of unmanned driving cars.

The unmanned driving taxis help to give the public access to quality shared mobility services at a lower cost. In the Asian markets such as China, India, and Singapore, carpooling is popular among a large number of people, which is closely related to the level of economic development to some extent. In the economically developed European countries such as Britain, France, Germany, and other European markets, people's acceptance of carpooling is relatively low. There are, however, some differences in the acceptance of carpooling by different age groups. In the carpooling willingness survey for the European market, 45% of the respondents under the age of 30 said that they were willing to share a taxi, only 22% of the people over 50 years of age and older people are willing to accept the carpooling of taxis.

The cost of an unmanned driving taxi has great influence on mass carpooling: when the cost of the unmanned taxis is $20, only 11% of respondents are willing to carpool; when the cost of an unmanned driving taxi is reduced by 50% to $10, the proportion of respondents' willing to carpool increases to 37%; when the cost of an unmanned driving taxi is further reduced by 75% to $5, the proportion of respondents' willing to carpool increases to 52%. In addition, many respondents to the survey actively asked about the presence of personal safety features such as glass partitions and cameras in the vehicle, and said that they would be more willing to carpool the unmanned driving taxis if it was done properly.

The reason for using smaller vehicles like taxis rather than large buses is that the smaller and more decentralized vehicles have a rather positive impact on meeting individual travel needs, improving road smoothness, disease prevention and control, and security against terrorism.

Some people say that the bus environment may promote human interaction, but the fact is that, except for a few cases such as business meetings and trips with friends and relatives, people travel more for their own amusement and relaxation than to chat with strangers with whom they have no interaction. Therefore, large bus scenarios such as large buses not only do not promote interpersonal interactions, but also affect people's travel experience.

In the future, as the level of urban development continues to increase and the cost of the unmanned driving travel is reduced, more and more people will choose to carpool or ride independently in small unmanned driving vehicles.

12.4.3 *Intelligent traffic management based on the integration of vehicles and roads*

The exploration of intelligent traffic management model from the perspective of travel mode will meet the concept of people-oriented development. However, it is far from enough to consider only consumers if a sound intelligent traffic management model is to be established, and the government departments, manufacturers, dealers, operators, and other subjects involved also need to be considered. From the development and design of the unmanned driving cars to putting them into operation itself is a large and complex system project that requires collaboration and interaction among all parties.

The communication and interaction between cars and cars, cars and roads, and cars and people need to have a unified industry standard. At the same time, it is far enough to explore the in-depth potential value of big traffic data only on the basis of one Internet travel platform or car enterprise, and it requires the sharing and in-depth cooperation of the enterprise resource. In addition, the vehicle quality, traffic accidents, and other issues have not yet been clearly defined, and the more vociferous point of view is to make the car enterprises assume the responsibility for the entire process of the unmanned driving vehicles' manufacturing and road operations as one.

In this case, the car company will be responsible for the research, development, and configuration of the vehicle, and road construction and management, which will ultimately establish a set of car and road integration, integrated operation of the intelligent transportation industry model, to provide strong support for urban transportation development.

For the research, development and production of the unmanned driving cars, it is necessary to actively expand to the upstream and downstream of the industrial chain, which will help improve the value in the creation capacity of the industrial chain and reduce costs and risks. Baidu gives full play to its resource integration advantages to achieve in-depth cooperation with the government departments, parts manufacturers, travel service providers, vehicle manufacturers, industry associations, and to establish and improve the "smart travel" ecology, so that the general public may enjoy more convenient and efficient transportation services.

In August 2016, California Silicon Valley technology company Velodyne announced that its lidar company Velodyne LiDAR received investment from Baidu and Ford, with total investment up to $150 million. The addition of the capital has significantly reduced the price of lidar products, and thanks to Velodyne LiDAR's strong supply capacity, the supply price of Velodyne's 64-wire lidar, compared to the previous figure of 700,000 yuan, has been reduced to 500,000 yuan. A spokesperson for

Velodyne LiDAR noted that when the 64-wire lidar orders reach the million level, its cost will be reduced to $500.

In other words, the price is expected to cease to be a major obstacle limiting the sale of lidar, and the commercialization process of the unmanned driving technology will be further accelerated. Currently, many traditional car companies are investing in the unmanned driving startups in order to seize the limelight. In order to ensure the safety of the unmanned driving cars and enhance their control capabilities, in the future, the supply of the unmanned driving car services and operation and management of a certain area will be the responsibility of a single enterprise. In this case, the target customers that the unmanned driving car companies strive for will be no longer the individual consumers, but cities, regions, and even countries. From the perspective of industry competition, this will help reduce the industry's internal depletion, and the establishment of external competition and industry-based Inter-regional competition dominates the industry competition pattern.

The automated driving tools such as automated taxis are only one of many travel methods and people can also choose bicycles, electric vehicles, buses, subways, and other travel methods. The mass application of mobile Internet and mobile smart terminals are widespread, so that cities and regions with more diversified services and better infrastructure will attract more talented people.

From a technical point of view, it is feasible that the automatic driving safety can be improved to the level of large-scale applications, but after reaching a certain level, it is quite difficult to improve safety from 80% to 99% and then it is even more difficult to improve from 99% to 99.99%. It's true that the data analysis capabilities of big data technology and the computing power of cloud computing technology can surpass the humans', but new problems are emerging all the time. The reasonable resolution to the problem requires comprehensive consideration of emotional, moral, ethical, legal, and other factors, and the solutions cannot be designed only from the technical point of view.

In the long run, after the unmanned driving enters stage of commercialization, for quite a long time, we cannot leave the safety and security entirely to the unmanned driving cars and systems, but they should be combined with more reliable and safe auxiliary means such as classified car and travel on separate roads.

12.4.3.1 *Classifying vehicles into different uses*

The classification of vehicles into different uses will be developed and designed with differentiated design and operation management for the unmanned driving cars based on the demand characteristics of actual scenarios. From different perspectives, there are several ways to classify the unmanned driving vehicles.

(1) From the point of view of the driving range, the unmanned driving vehicles can be divided into short-distance vehicles that operate at low speeds within cities and long-distance vehicles that operate at high speeds between cities.
(2) From the point of view of loading objects, the unmanned driving vehicles can be divided into passenger cars that serve people and trucks that serve freight.
(3) From the point of view of the number of passengers, the unmanned driving vehicles can be divided into single-passenger, double-passenger, and multi-passenger vehicles.
(4) From the point of view of different users, the unmanned driving vehicles can be divided into police cars, commercial vehicles, business vehicles, and family cars.
(5) From the point of view of special scenes, the unmanned driving vehicles can be divided into emergency vehicles, fire trucks, energy supply vehicles, and engineering emergency vehicles.

The application of vehicles according to the classification of vehicles is of great value in reducing resource waste, controlling costs, and improving service quality. For example, those short distance vehicles running at low speed inside the city only need to update the local map in real time, which can help retain more detailed local traffic information, reduce the total amount of data, and improve the speed of the vehicle system; for indoor short distance vehicles and intercity vehicles, different design solutions can be adopted to differentiate tires, lighting, body materials, sensors, etc., to reduce resource waste and improve the utilization of resources.

After the definition of the vehicle operating zone, the scale of vehicles in the region will be effectively controlled, which can improve the efficiency and quality of traffic operation and reduce traffic congestion and traffic accidents. With the popularization of mobile payment and the reform of toll policy, the road toll stations will cease to exist in the future and they will be replaced by relay stations that serve users for transferring passengers and goods.

In the future, there will be no need for passengers or luggage to get off the car to complete the transfer or transit. For example, passengers will be able to get on and off the high-speed train without stopping when the high-speed train reaches the station. For those traveling in the unmanned driving cars, when the road ahead is impassable due to traffic accidents or congestion, passengers traveling in the opposite direction will not have to turn back in the same direction and can continue to get to their destination by exchanging vehicles.

12.4.3.2 *Travel on separate roads*

Travel on separate roads will set different lanes for pedestrians, bicycles, electric vehicles, private cars, buses, and other different vehicles and they will travel on the corresponding lanes and control their travel routes within the travel range. At present, the highway and rail transportation have almost realized the travel on separate roads. In the early application of the unmanned driving cars, it is necessary to set up special lanes for the unmanned driving cars in order to improve the safety and reduce the negative impact of accidents.

In the future, when the traffic lights at the traffic intersections are canceled, the lane separation from people to vehicles will be an inevitable choice, because it is difficult for ordinary people to judge the route of the unmanned driving vehicles, and at the same time, it is difficult for the unmanned driving vehicle systems to find fixed patterns in the large number of irregular flows. Indeed, with the widespread application of the Internet of Things technology, pedestrians will be able to interact with the unmanned driving cars in the future, but passengers in the cars may not be willing to compromise, which may lead to unnecessary conflicts and disputes, and lane splitting will effectively solve this problem.

In addition, when the unmanned driving cars and manned driving cars are working on the same road, due to the lack of binding force of the intelligent traffic center system with unified dispatching function on the human-driven cars, the instructions will not be effectively implemented, and conflicts would occur frequently, which is by no means the desired outcome.

12.5 Key Factors in Unmanned Driving from Concept to Practice

12.5.1 *Safety: The inevitable destination for unmanned driving development*

The quest for safety is spurring a rapid growth in demand for the unmanned driving vehicles. According to the data released by the World Health Organization, globally, approximately more than 1.2 million people die in road traffic accidents each year. In China, human-caused traffic accidents account for 93%, with an average of 500 people dying in traffic accidents every day. With the emergence of the unmanned driving cars, the traditional "human–car–road" ternary control will be transformed into the "car–road" binary control, with the result that the uncontrolled drivers will be eliminated and "zero violations" will be achieved so that the efficiency and safety of the traffic system will be greatly improved and the frequency of traffic accidents will be significantly reduced, thus driving safety will be effectively ensured. Studies have shown that after automobiles are equipped with automatic braking systems, the incidence of rear-end accidents will be reduced by 40%; if all motor vehicles are equipped with the automatic braking systems, the number of traffic accidents can be reduced by 700,000 per year.

The premise of the unmanned driving development is safety. The safety and reliability of unmanned driving has not yet been widely recognized by the people because unmanned driving technology is still at the trial and improvement stage. For example, the British regulators require that the unmanned driving cars have to be monitored by humans when they are driven on the road, and they have to be able to be switched to the manned driving mode at any time; Germany plans to release a series of regulatory policies on automated vehicles, including the requirement that the unmanned driving cars have to be equipped with a steering wheel, have a device installed that is similar to the black box of civil aircraft so that in the event of a car accident the relevant investigators can be based on the black box for the causes of the accident analysis, therefore the responsibility of all parties will be made clear.

Volvo in Sweden has conducted a survey which showed that more than half of the consumers support the installation of steering wheels in automatic cars, and 80% of the consumers said that the manufacturer should be responsible for settling claims if an accident occurred in a self-driving car because the cause of the accident was a malfunctioning autopilot system. In fact, if an unmanned driving car encounters an emergency while driving, the user does not trust the autopilot system to respond effectively and the struggle for control is likely to bring about extremely serious consequences. This is because the drivers tend to be more relaxed during the process of autopilot and do not pay much attention to the driving conditions, they are likely to make wrong judgments and put themselves in danger when faced with unexpected situations.

When faced with such a situation (a traffic accident) where the driver and vehicle are competing for control, it is difficult for the traffic department to decide which is the party responsible for the accident. On the other hand, if the driver is required to remain vigilant and ready to switch to human driving mode while the car is in motion, it would be better to adopt human driving directly. This "human or car" design makes the unmanned driving meaningless and increases the burden on the driver's mind, which will make the desire to drive freely and comfortably futile.

Moreover, in essence, this "human or car" design is an irresponsible manifestation of the safety of unmanned driving. If there is a "back road" of manned driving, the relevant personnel will not focus on the unmanned driving safety issues, and the unified standards for the design and manufacture of the unmanned vehicles cannot be built in a short period of time, therefore the user's safety will not be effectively guaranteed.

On May 7, 2016, the Tesla Autopilot that activated the automatic driving assistance system collided with a white tractor-trailer truck while driving, the top of the Autopilot was completely destroyed and the driver died on the spot. Tesla's explanation for the accident is that the white truck was not recognized against a blue sky background and the side of the tractor-trailer truck was suspended on the ground from Autopilot's viewpoint, which caused the Autopilot system to miscalculate and fail to activate the automatic brakes, resulting in a collision between the two vehicles. Also, because the location of the vehicle collision was at the windshield of Autopilot, the crash safety system failed to work, killing the driver at the scene.

The accident caused an uproar because Tesla had claimed that a fatal traffic accident occurs every 94 million miles in the US and that the Autopilot-assisted driving system had traveled 130 million miles at the time of the crash. Also, Tesla stressed that the unmanned driving technology is constantly evolving and Autopilot is not perfect. Therefore, for this accident, who should exactly be blamed? Some believe that the responsible party is Tesla, while others believe that the responsible party is the inattentive driver, and still others believe that the responsible party is the sensor supplier Mobileye. This perception of the unmanned driving traffic accident reminded us of the alternating current battle between Tesla and Edison.

In 1880, Tesla invented the world's first alternating current generator and Edison invented the electric chair in order to bring awareness of the dangers of alternating current to people. Later, after a fierce battle between the two, Tesla's alternating current prevailed and made an outstanding contribution to the development of human civilization. Thus, all new things are a symbiosis of pros and cons, and there is no absolute good or bad. Therefore, in the face of new things — unmanned driving cars — we should grasp the trend of development and we cannot completely deny it because of an accident.

So, how to solve the issue of safety of the unmanned driving cars?

In order to reduce the occurrence of traffic accidents in the course of driving unmanned vehicles, the unmanned vehicle manufacturers should establish a strict system of regular return of vehicles to the factory for maintenance and driving emergency disposal service system, especially to establish a "dual system" control mode to solve the problem of possible sabotage of the vehicle network, such as hacker intrusion, virus damage, and so on. The primary system in "dual system" refers to the open system of the networking, while the secondary system refers to the independent closed system. Once the driver discovers that the primary system has been sabotaged, the driver can immediately switch to the secondary system to ensure that the vehicle runs normally and that the driver and passengers get off safely, and then the vehicle needs to be sent to a maintenance organization for repair.

If the driver does not find any abnormality when the primary system has been damaged, for example, the car will collide onto the obstacle in front instead of turning for a roundabout driving because the code of the system was revised and the driver may not be able to switch the systems in time, which will result in the traffic accident. In other words, the vehicle manufacturers must install protective software for the unmanned driving cars so that the vehicle can automatically detect faults, check and kill viruses, and take responsibility for the traffic accidents caused by improper protection, so as not to shirk the responsibility of vehicle manufacturers in the event of traffic accidents, resulting in damage to the rights and interests of the users. In addition, the unmanned vehicle manufacturers should also ensure that the vehicle's internal and external interfaces are difficult to dismantle and change to prevent users and others from artificially destroying the vehicle, thereby the safety performance of the vehicle is to be effectively improved.

In addition, if the transportation department stipulates that the unmanned driving vehicles must be able to switch to human driving mode at any time, then the public transportation companies and taxi companies that are best suited to introduce the unmanned driving vehicles will completely lose the advantage of being free of human labor. The reason is that if the companies do not provide the unmanned driving cars with full-time drivers, they must first check the user's driving license, whether they drink alcohol, or take drugs, etc. The whole process is not simplified, but it becomes more and more cumbersome, and the unmanned driving car's purpose of convenience may be dashed.

In summary, the traffic departments should not make the regulations that "the unmanned driving cars must be able to switch to artificial driving mode at any time". It will otherwise lead to unclear standards and unclear responsibility in the entire industry, which will bring the development of the industry into a confusing, embarrassing situation. In other words, there is a major loophole in the unmanned driving if the practice of human assistance is accepted in order to avoid traffic accidents. To achieve the ultimate goal of the unmanned driving development, the transportation authorities must reform the current transportation system and clarify the responsibilities of the unmanned driving vehicle manufacturers, the unmanned driving system software providers, and the unmanned driving bus companies as well as the liability of insurance companies so that the users can ride the unmanned cars with confidence. Recently, the US federal government and some states have affirmed the legality of the unmanned driving and Google unmanned driving car has been affirmed by the US Highway Safety Administration, and it confirmed that the vehicle accepts the control of the automatic driving system, rather than the owner's manual control. In other words, the previous rule that "the unmanned driving cars must be equipped with a steering wheel and brakes" is no longer valid.

In the future, unmanned driving cars will no longer be equipped with steering wheels, pedals, instruments, levers, observation mirrors, and other equipment, instead they will be equipped with panoramic doors and windows, 3D screens, spacious open carriages, mobile communication and office equipment, etc. By then, the car will no longer be the "connection" to life, but will become an extension of the living space, and people's time and energy will be completely liberated to put into more meaningful and valuable things.

12.5.2 *Performance: The unmanned driving practice*

As the users' demand for unmanned cars explodes, unmanned driving cars will quickly enter people's lives and become an important part of their daily living spaces, and they will become the most important means of transportation. At that time, the functioning of the pure electric unmanned driving car will be better and better. Unmanned driving cars will drive not only on the road, but also directly into the office building, residential buildings, and even into the rooms, they would become the user's office, lounge, meeting room, and they may become a fully functional "caravan".

What's more, the offices and residential areas may also evolve into giant "caravans". The unmanned driving cars will be nested in layers like wrappings, with small cars being set into medium and medium cars into big cars, giving rise to infinite changes. Of course, the purely electric unmanned driving cars could also be parked in the open and charged by using solar energy while they continue to provide services to users besides driving.

The car of the new concept will draw on the idea of the bumper cars and be equipped with flexible environment-friendly materials with strong toughness for its body in order to reduce friction and collision and buffer the damage caused by any friction, collision, and accidental impact. The traditional metal body is sturdy and beautiful, but it is poor in terms of environmental protection and energy saving, and it is not resistant to corrosion, is poor in sound insulation, poor in safety, and not very practical. The use of new flexible materials for the vehicle body can give the body many changes, which will create a variety of models such as stretching cars, deformation vehicles, and amphibious vehicles on land, in water, and in the air.

In addition, using flexible materials to make the vehicle body is to make the body storage, assembly, and transportation more convenient, and the relatively simple structure of the pure unmanned driving electric vehicles is comparatively simple, which will make vehicle manufacturing become easier so that the costs will drop significantly. It is conceivable that in the near future there will be unmanned driving electric vehicles of different appearances and sizes on roads all over the world. The unmanned driving vehicles can be comparable to, or even more popular than, the traditional housing with their spacious cabins, rich and perfect functions, closer to nature, lower prices, greater flexibility, and disaster resistance.

Furthermore, the unmanned driving cars also call for energy convenience. For example, a pure unmanned driving electric car can find a charging station to finish charging ahead of time without human control, or use their free time to charge, and the occupants can use the charging time to deal with other matters, so even if a pure unmanned driving electric car is not currently able to be charged fast, there is no need to worry about the user's wasting time. In the case of unmanned driving cars' feeding of fuel, the vehicle will automatically go to a gas station for refueling without the presence of the passenger in the car, which can also reduce the safety risks. In addition, in an unmanned driving environment, the charging stations can be set up with fewer charging stations to reduce infrastructure construction and maintenance costs.

In terms of range, although the range of a pure unmanned driving electric vehicle is poorer than that of a gasoline vehicle, they can complete mobile charging while driving because the pure unmanned driving electric vehicles have higher energy security and better operating capability, which is just like an airplane refueling in air. The pure unmanned driving electric vehicle that needs to perform mobile charging can turn to a dedicated charging service vehicle, or to other pure unmanned driving electric vehicles that are traveling with them.

The methods for charging the unmanned driving car can be the methods of replacing the battery directly or adopting vehicle-to-vehicle docking. Regardless of the charging method, charging an unmanned driving car is much simpler than charging a manned driving car. It is mainly private cars that carry out mobile charging on the way, while buses and taxis can calculate the distance traveled and the amount of electricity, and arrange for passengers to change the car to travel at the right time.

12.5.3 *Experience: Meeting the user's lifestyle and entertainment needs*

12.5.3.1 *The optimization of appearance and styling, space layout, and basic functions*

Under the unmanned driving conditions, the driver's body posture can be more relaxed, not only in a sitting position, but also in a lying or standing position, etc. In order to respond to the driver's needs, the carriage must be raised and enlarged, and the seats must be stowed and retracted, either as seats for the driver to recline, or flattened into a bed for the driver to lie down, or even folded up to allow the driver to stand freely. The elevated and widened doors can be of a "walk-in" design, which is much more convenient than the "sit-in" design of traditional sedans. In addition, a small folding table can be placed inside the car to create a simple office space.

The unmanned driving car manufacturers can design a variety of models, such as single-person, double-person, and multi-person vehicles in accordance with the users' needs in order to save energy and reduce travel costs. In fact, the single-person unmanned driving cars are bound to be very popular among the office workers. The single unmanned driving car can be designed as a single row of wide-wheeled vehicle with an appearance similar to a large motorcycle with a canopy, which not only saves space and is more flexible in driving, but also saves energy and is environmentally friendly as well. The multi-person unmanned driving vehicles can be further divided into buses that can share space, mid-buses and pods that have privacy on space, which will further enhance the driving safety. The unmanned driving cars can be assembled like trains to form a large vehicle that is connected to the internal space because the unmanned driving vehicles are designed according to a unified standard. If there are more items to be transported, other unmanned vehicles can be enlisted to help by combining them together to expand the space inside the vehicle.

Moreover, unmanned driving vehicles can also be divided into short-distance and long-distance vehicles according to the driving environment and tasks, and the short-distance unmanned driving vehicles mainly travel within the city and should not be too fast or complex in function, instead they should have a strong integrated perception ability. Long-distance vehicles, however, mainly travel on intercity highways, so the speed can be designed to be high, the functions should be appropriately rich, and the integrated perception ability can be slightly weakened. This vehicle division of labor can save resources and reduce costs.

12.5.3.2 *Providing communication office and life entertainment services based on the internet of vehicles*

In an unmanned driving environment, the drivers do not have to drive by themselves and they can use their travel time for other things. According to the estimation of the US Department of Transportation, each person spends about 52 minutes a day commuting to and from work. In an unmanned driving car, people can make use of that time more efficiently. If the carriage of an unmanned driving car is large enough and the travel time is long enough, and if it is equipped with the necessary information technology equipment, the unmanned driving vehicles can create a comfortable office space and make mobile office and mobile home a reality, which will truly realize the idea that makes the "car as a service" and "car as a life".

In this way, the unmanned driving cars would also get the users' privacy involved, and in the case of the public officials, may get the national security involved as well. The unmanned driving cars can make use of the non-real-name payment methods in order to protect privacy and national security. At the same time, the users can turn off the location function of their phones, do not have to authenticate their real names when using the mobile internet, and even disconnect the internet connection during the ride. After these actions are taken, even if the vehicle is still under the surveillance of Internet of Vehicles, there will be no real-name correspondence between the vehicle and the user, therefore the user's privacy can be greatly guaranteed. In addition, the user can manually turn off the camera and pointing device in the car during the journey to avoid leakage of privacy.

Of course, the user is blocking security surveillance by taking these measures, which will prevent the traffic police from dealing with accidents quickly in case of accidents. The unmanned driving car will become a "black car" in the true sense of the word after cutting off the network connection, it can neither receive the command sent by the command center, nor can the driving intention be fed back to the command center. The command center can only monitor it through the monitoring on both sides of the road and the information provided by the surrounding vehicles. In order to ensure the safety of the vehicle, the command center and management department must reserve certain monitoring rights for buses, taxis, and other modes of transport, and make use of automatic detection technology to detect dangerous and lost items in a timely manner to ensure the safety of passengers and provide convenience for passengers to retrieve their belongings.

12.6 Application of the Unmanned Driving Technology in Urban Rail Transit

12.6.1 *Profile of the application of domestic and foreign metro automation technology*

With the continuous improvement of China's economic development level, the construction process of rail transit systems in various areas has been accelerated. With the active efforts of the universities, research institutions, enterprises, and other parties, new urban rail transit technology has emerged in large numbers, especially the application of fully automatic unmanned driving technology in the field of urban rail transit, which is of great practical significance to promote the quality and efficiency of urban rail transit systems, improve the urban rail transit network, and dock with the rail transit system standards of traffic powerhouses.

12.6.1.1 *Advantages of unmanned driving technology*

(1) The fully automated unmanned driving subway can automatically wake up, be dormant, clean, stop, drive, open and close the doors, repair faults, etc. The different modes of operation such as regular operation, downgraded operation, interrupted operation, etc., can be set up to meet the individual needs in different scenarios and improve operational safety and ride experience.
(2) The unmanned driving trains can be started and stopped more evenly, which can therefore bring passengers a high-quality travel experience at faster speed and reduce people's travel costs.
(3) The fully automated unmanned driving subway has a higher level of automation, and it can significantly reduce the cost of manpower and material resources. It is true that the construction of this kind of subway may cost more, but its positive role in improving the efficiency and quality of subway operation and reducing operating costs is obvious.

12.6.1.2 *Profile of the application of foreign unmanned driving technology in urban rail transit*

(1) Metro Line 1 in Paris: The line was opened as early as July 1900, and it is the busiest and oldest east-west line in the region, with a total length of 17 km covering 25 stations. The Metro covers such famous attractions as the Arc de Triomphe, the Louvre, Charles de Gaulle Square, and the Champs Elysees, etc. In 2011, the line applied the unmanned driving trains based on the Trainguard MT safety and control system.
(2) The Barcelona Metro Line 9 in Spain: The line is 41.4 km long and its real-time data transmission is achieved by the extended spectrum radio, and the train is driven automatically by a fully automated unmanned system.
(3) In Aichi Prefecture, Nagoya, Japan, a set of HSST-100 magnetic levitation system was built, the line is 8.9 km long covering 9 platforms, and it makes use of ATO (Automatic Train Operation) unmanned operation mode.

12.6.1.3 *Profile of the unmanned driving technology applications in China*

There are already a significant number of subway lines in China which make full use of the automated unmanned driving technology.

(1) Beijing Rail Transit Airport Line: This line adopts fully automatic unmanned driving trains, which can optimize and adjust the train running strategy and running density according to the passenger flow, and at the same time, it is equipped with a relatively complete operation assistance system that can locate the trains in real time, make the trains fold back automatically, shorten the running interval, and improve the passenger carrying capacity.
(2) Shanghai Metro Line 10: This line is the first line in China to apply the unmanned driving technology which can drive with more accurate acceleration and deceleration control, with no need to consider the driver's shift plan, and a large number of self-checks after starting up the train every day. The train has a higher level of safety and reliability.
(3) Beijing Rail Transit Yanfang Line: This line is the first fully automatic running metro line developed independently in China and the second

line applying the unmanned driving technology following Shanghai Metro Line 10. The first phase opened at the end of 2017 and the second phase was expected to open in 2020, with trains running smoothly and efficiently, which can provide passengers with quality riding experience and reduce energy consumption by 10% to 15% compared with the ordinary trains.

Much convenience had been brought by the application of the unmanned driving technology thanks to the urban rail transit running on a specific track, and less interference from pedestrians and other vehicles is involved. The research on and development of the unmanned driving technology will not only promote the increasing improvement of urban rail transit, but also lend valuable reference experience for the large-scale application of the unmanned driving technology in the field of transportation so that human society will usher in a new chapter of the era of the unmanned transportation.

12.6.2 *Functional features of unmanned urban rail transit system*

In recent years, the development process of urban rail transit has been accelerating, and the corresponding supporting technology has also been rapidly updated. Many first-tier cities with large populations have to bear greater pressure on passenger flow, therefore they can only shorten the process of new line construction and increase the number of busy lines and frequency of shuttles, which must absorb more crew members, drivers, etc., and undoubtedly more costs have to be invested at the same time.

In addition, increasing the frequency of trains will increase the labor intensity of manual work, and if heavy physical labor is maintained for a long time, safety accidents are likely to occur due to fatigue. Considering this situation, the urban rail transit industry is increasingly involved with the unmanned driving technology.

In summary, the functional features of the unmanned driving systems in urban rail transit are reflected in the following aspects:

(1) The automated control over the train's operation is specifically displayed in the train starting, train alignment, obstacle recognition, train and screen door monitoring, formation adjustment, and many other

working processes. The operation of the train can be intelligent operation instead of the traditional manual operation, which achieves electro-mechanical integration of operation.

(2) The efficiency and comfort of the trains can be improved and energy can be saved as well. The goal of saving energy can be achieved through taking the planned timetable as a reference to control the train's running interval and running time reasonably; and the impact rate of the traction brake is effectively controlled by computer and the acceleration and deceleration are only carried out when necessary. The changes to the opening scheme are made and the reasonableness of the train running interval is maintained on the basis of the analysis of historical passenger space-time distribution, real-time cross-sectional passenger flow changes, large passenger flow warning information by combining them with the original plan, and if it is necessary, the passengers will be immediately evacuated.
(3) The operation of the vehicle, communication, and signal systems can cooperate with each other and form relatively perfect subsystems in each part, which gives full play to the role of the Integrated Supervisory Control System (ISCS). The connection between different subsystems is strengthened to exert different functions in train rescue, equipment trouble shooting, and other work. The COCC, combined with ISCS, is used to jointly implement automated network operations and improve operational efficiency as a whole.
(4) The human–machine monitoring interface and the corresponding interface is opened to send relevant information for operators, passengers, and maintenance departments when the trains remain in normal running condition, so as to make it convenient for the demanders to inquire and use; if abnormalities occur, the safety program will be automatically enabled and the early warning will be given so as to clarify the degree of influence and find the root cause of the problem in an automated way or in cooperation with technical personnel, and the interface service is opened to facilitate the intervention of maintenance personnel.

12.6.3 *The requirements of the unmanned driving for urban rail transit*

(1) Listing and control systems require high safety. As an important part of the unmanned driving system, the control system is responsible for

a wide range of safety control, combining a variety of functions in coordination with the manned system. The relevant problems are to be solved on the basis of the implementation of safety analysis in order to improve the safety of the overall operation. For example, in the process of designing the line, it is necessary to ensure that the driving environment is not disturbed by external factors, and in the process of subsequent development, it is necessary to continuously improve the safety system and timely perceive safety risks through sensors.

(2) Only with sufficient safety, maintainability, reliability, and availability can the normal operation be maintained due to the limited number of humans involved in an unmanned system. In case of equipment failure or emergency, it is necessary to use the remote monitoring system to respond to the problem in a timely and effective manner; in addition, the unmanned system should also have the ability to analyze scenarios and issue emergency plans in a timely manner. In the event of an abnormal situation, the formulation of the plan should comprehensively consider the different needs of passengers and put the protection of passengers' personal safety as the first priority.

(3) CBTC (Communication-Based Train Control) mobile occlusion system is widely used in the unmanned driving system, and the most important thing is to use a sound network to maintain the normal operation of the system. The value of a sound network system can be highlighted even more if the synergy between different systems is to be exploited in the operation process.

(4) The unmanned driving systems are a subversion of traditional manual management methods, which requires the urban rail transit departments to change their existing construction methods, adjust their organizational structure, actively learn from the excellent experience of successful projects at home and abroad, and formulate more reasonable planning schemes on the basis of the specific and special scenarios of China's cities to promote the development and improvement of the unmanned driving systems from all aspects and improve the management capabilities of relevant departments.

(5) The construction of unmanned driving system can promote the development of urban rail transit system from many aspects and improve its level of automation and intelligence. In addition to replacing driver operation with automated operation, the value of the unmanned driving systems is also reflected at the dominant level, which should have the courage to highlight the constraints of traditional concepts and

actively innovate. In the subsequent development process, the unmanned driving systems will focus on the response to problems in passenger flow, operations, troubleshooting, and disaster management by using big data and artificial intelligence technology to improve operational strategies and providing quality services to passengers and other participants in transport operations.

12.6.4 *Difficulties in the application of the unmanned driving in urban rail transit*

(1) Online Testing and Monitoring: Monitoring vehicle operating status is only part of the online monitoring work of the unmanned driving system and there are a number of other systems including the automatic booking system, signaling system, passenger information system, fire alarm system, etc. In the event of system failure, it should also respond in a timely manner and make the corresponding resolution. Obstacle detection is an essential part of the unmanned driving system that cannot be ignored, and in addition to the detection, appropriate handling measures should be taken. Nowadays, contact detection is the commonly used detection method in the industry, and there is lack of effective non-contact obstacle detection. If an obstacle is detected, it will be responded by emergency braking, which makes it difficult to improve the efficiency of unmanned driving. At the same time, it is necessary to focus on the pyrotechnic detection, anti-terrorist riot monitoring, foreign object intrusion monitoring, and their corresponding handling methods to reduce the resistance in the operation of the unmanned driving vehicles.

(2) Automatic Linkage of Vehicles: If the train encounters an emergency situation during operation that causes an anchor, it is necessary to start an emergency rescue. In the process of emergency rescue, the broken-down train should be connected to the rescue vehicle, the automatic connection can solve this problem. However, if we want to quickly complete the one-time automatic linkage between the anchored train and the rescue vehicle at the rescue site, we have to improve the relevant technical levels. The current automatic linkage has many limitations, and only the communication trains that run in the same direction and have their doors in a closed and locked state can achieve the

automatic linkage. In the future, more attention should be paid to improve the rescue efficiency of broken-down trains and reduce the limitations of the automatic linkage.

(3) Emergency Accident Handling in Operation: A serious crisis can occur if an emergency occurs in operation, especially if an unforeseen situation occurs in the train. For example, a serious crisis may take place when a passenger presses the emergency brake button on a train which is in the operating zone and the train is in fully automatic operating mode without a following maintenance crew. For this kind of situation, the operating managers need to develop the targeted solutions. From time to time, there will be cases of passengers violating the rules of subway operation in the stations, for example, the passengers force their ways into booth screen doors or train doors being jammed while the train is closing the doors, which will result in the abnormal operation of the train. All these emergencies should be dealt with by the station staff, which requires them to predict the risks and make efficient resolutions.

At present, there are still many technical difficulties that need to be overcome in unmanned driving like the construction of the better signal systems. The subways of first-tier cities in China carry a lot of passenger pressure including the high frequency of departures and the demanding requirements for safety, but the function of the subway system will be gradually improved on the basis of increasing technical levels, and the difficulties faced by the unmanned driving technology will also gradually be overcome. When the development of the operating system is gradually bettered, more and more urban rail transit will introduce the unmanned driving technology, which will promote the metro operations to achieve a leapfrog development.

12.6.5 *Solutions of urban rail transit based on unmanned driving*

The urban rail transit, as an important part of urban transportation construction, is the mainstream tool for urban residents' daily travel, which has a crucial impact on their travel experience. In recent years, the urban rail transit technology ushered in a period of rapid development and a large number of unmanned driving technology applications emerged, which provide new ideas for controlling train operating costs and improving the quality of travel services, and it is a major hot direction in the field of transportation. Therefore, the analysis of the unmanned driving technology application in the urban rail transit to find effective solutions will promote this technology in the city, which will have a positive influence on the large-scale application of unmanned driving technology in the field of rail transportation.

12.6.5.1 *Guarantee of facilities and equipments*

Train Functional Guarantee: The unmanned driving train in operation does not need human involvement because the train not only has the automatic operating functions but also tests their breakdown. When there is a potential safety hazard or fault in a certain part of the train, it is necessary for the train to be able to find out the root of the problem, provide effective solutions, take effective measures in time to prevent the situation from worsening so that the stable and safe operation of the train can be ensured.

It is necessary for the train to be able to automatically perform backward positioning. The execution of loading and unloading operations is likely to cause safety accidents when the unmanned driving train crosses the stopping position. Therefore, it is necessary to ensure that the unmanned driving train can perform backward positioning to ensure the safety of passengers on and off the train.

It is also necessary for the train to obtain the function of detecting and handling obstacles automatically. Major traffic accidents may happen when the unmanned driving train encounters obstacles in the course of operation and no effective measures are taken in a timely manner. Therefore, the unmanned driving trains must be able to automatically detect and effectively handle obstacles in order to ensure the safety of trains in operation.

It is necessary for the train to obtain the function of video monitoring as well. With the support of the video monitoring system, the train operation control center can keep track of the internal and external conditions of the train in real time, so that the control center can make timely and effective adjustments to prevent traffic accidents when problems arise in the operation of the train.

Signal: The signal is an important factor that affects the operation of unmanned driving trains. Only when the signal system is running stably can the train complete acceleration, deceleration, turning, stopping, opening and closing the doors, etc. Safety accidents such as doors pinching passengers and trains derailing may happen when there are problems with the signal system. Therefore, the automatic detection and maintenance of the signal system is particularly critical.

Evacuation Platform: The handling of accidents may be more challenging when the accidents happen while the trains are in operation because the staff cannot get to the spot to handle the accidents effectively. In order to reduce the negative impact of the accidents, an evacuation platform has to be set up in the tunnel so that passengers can be moved to a safe area if an accident occurs.

The unmanned driving technology itself has a high dependence on the supporting hardware and software facilities. The solutions have to be made manually when the supporting hardware and software facilities have problems and the control center cannot effectively solve the problem. Therefore, the train operation department must carry out real-time inspection of the train supporting facilities and equipment to fully ensure the reliability of unmanned driving train operations.

12.6.5.2 *Innovative management model*

In some cases, the remote control from the control center cannot effectively solve the problem and staff must enter the interior of the train to deal with it when there are problems with train in operation. The manual handling also requires a certain amount of time. The problems cannot be handled in a timely manner by the crew as is generally the case with traditional human-piloted trains. Therefore, it is necessary to make innovation in management mode in order to improve the processing speed of the unmanned driving trains.

The driver's role does not exist in the unmanned driving trains, and the driver's functions in the traditional manned driving mode will be transferred to the dispatching department, which requires the latter to be more refined, that is, the dispatching department of the unmanned driving trains should have strong communication and coordination skills and professional driving skills, etc., so that it can monitor train passenger flow in real time, respond immediately to feedback from train passengers, and provide passengers with effective solutions to problems in the shortest possible time.

In addition, a more complete maintenance and dispatch system should be established in order to solve more complex unmanned train operation problems so that in the event of complex problems such as signaling system failures and train track damage, the problems can be solved quickly. To this end, the maintenance dispatching, traffic dispatching, and passenger dispatching need to maintain efficient coordination, which requires the integration of all three into the control center dispatching system at the same time.

12.6.5.3 *Personnel Quality*

The application of the unmanned driving technology in the field of urban rail transit has a high dependence on the professionals. Although there may be no longer staff inside the train, this does not mean that the unmanned driving trains no longer need human support. On the contrary, there is a higher demand for the overall quality and skills of the relevant staff.

(1) Control Center: For example, when a train malfunction occurs, it is more likely to be handled by remote control from a control center, whereas it may have previously been handled simply by the driver. Now since there is no crew on the unmanned driving train, the train dispatcher should not only have dispatching skills, but also train incident handling skills such as calming down the passengers and analyzing and resolving train malfunctions. Of course, in order for the train dispatchers to have these skills, they need to be professionally trained.
(2) Stations: With the application of the unmanned driving technology to the urban rail transit, a greater number of people with greater expertise and higher service levels need to be staffed at the stations in order to respond effectively to all kinds of accidents that may occur with trains.

12.6.5.4 *Passengers' cooperation*

Passengers are required to do a good job in cooperating with the unmanned driving trains when the trains are in operation. On the one hand, the human interference in train operation should be reduced. If part of the passengers' interference in the train operation is not stopped in a timely manner, it will have great negative influence on the train operation because there are no crew on the train. In this case, the passengers' initiative in cooperation is required besides the severe punishment for the interference actions on the train operation. On the other hand, the passengers may offer assistance in the rescue. After a train accident, it takes some time for the professional staff to arrive at the scene of the accident and they may miss the best time to rescue. If the passengers can spontaneously handle the accident appropriately by following the accident handling requirements, they will effectively reduce the hazards of the accident.

The prospect of the application of the unmanned driving technology to the rail transit transportation deserves our great expectation. It is the government departments that need to take a balanced approach, grasp the key points, guide the safe and standardized application and development of the unmanned driving technology while encouraging innovation and providing a more excellent business environment for entrepreneurs and enterprises in order to ensure the rapid and stable development of the unmanned driving technology in China.

References

[1] Lu Guangming. Ningbo Builds Intelligent Transportation Management System. *People's Public Security Daily, Traffic Safety Weekly*, August 18, 2017, p. 2.

[2] Li Haifeng and Ma Xiaolei. "Youth Mainstay" of Transportation Big Data. *Science and Technology Daily*, October 10, 2017, p. 6.

[3] Liu Qi. Big Data Tells You Why the Traffic Accident Happened. *Changsha Evening News*, April 8, 2017, p. 7.

[4] Lin Gang. Qingdao Intelligent Parking Comprehensive Platform was Launched. *Qingdao Daily*, July 27, 2017, p. 4.

[5] Luo Yunhui, Li Lin, and Qi Wenzhou. Research on Single Point Traffic Signal Timing Optimization Strategy Based on Big Data. *Highway and Automobile Transportation*, 2017 (04): 22–27.

[6] Xu Jihua, Feng Qina, and Chen Zhenru. *Smart Government: The Coming of the Era of Big Data Governance*, Beijing: CITIC Press, 2014.

[7] Zhang Xiaoming. "War" Heat, Not just Physical Work. *Wen Wei Po*, July 13, 2017, p. 2.

[8] Li Kun. Build a World Intelligent Platform to Help the Coordinated Development of Beijing, Tianjin and Hebei. *China Reform News*, July 3, 2017, p. 7.

[9] Yan Wei. Research on Transportation Big Data and Application Technology. Tianjin Electronic Industry Association. *Proceedings of 2017 Annual Meeting of Tianjin Electronic Industry Association*. Tianjin Electronic Industry Association: 2017, p. 4.

[10] Xu Zongben. Academician of Chinese Academy of Sciences, Professor of Xi'an Jiaotong University, Using Big Data Well Requires Great Wisdom. *China Education Network*, 2017 (06): 31–32.

[11] Hu Jihua, Gao Lixiao, and Liang Jiaxian. OD Matrix Inference Method of Bus Routes Based on Transportation Big Data. *Science Technology and Engineering*, 2017 (11): 309–314.
[12] 7its.com. Wechat Official Platform: The Current Situation, Application and Benefits of the Construction of Changsha Intelligent Transportation System, November 30, 2017.
[13] Zhao Bingyu. Meet the Big Transportation in the Era of Big Data. *Yan'an Daily*, July 17, 2017, p. 1.
[14] Yu Shuo and Li Zeyu. Research on Transportation Big Data and Application Technology. *China High Technology Enterprises*, 2017 (04): 90–91.
[15] Xia Huan, Editor in Chief. *Collection of Smart City Industry Solutions Centered on Data*, Wuhan: China University of Geosciences Press (2016), p. 11.
[16] Xiao Ziqian, Chen Jingyou, and Fu Shi. Overview of the Development of Intelligent Transportation System in the Context of Big Data. *Software Guide*, 2017 (01): 182–184.
[17] Zhao Guanghui. Thoughts on the Development of Big-data Transportation in the Context of "Internet +". *Logistics Technology*, 2016 (06): 19–24.
[18] Zhang Bin, Mao Lin, and Zhang Yiwen. Research on the Application Relevance of Traffic Big Data based on Real-time Road Condition. *Henan Science and Technology*, 2016 (11): 108–110.
[19] Yan Junwei, Ling Weiqing, and Wang Jian. An Ontology Based Transportation Big Data Analysis Framework. *Computer Knowledge and Technology*, 2016 (01): 25–27.
[20] Zhang Hong, Wang Xiaoming, Guo Xiucheng, Cao Jie, Zhu Xusheng, and Guo Yirong. Application of GPS Trajectory Big Data of Taxi in Intelligent Transportation. *Journal of Lanzhou University of Technology*, 2016 (01): 109–114.
[21] Chen Tao. Research on the Application of Big Data in Intelligent Transportation System. *Intelligent City*, 2016 (02): 36–37.
[22] Chen Ran. Analysis Framework of Transportation Big Data Based on Ontology. *Technology and Economic Guide*, 2016 (06): 27.
[23] Zhang Zi. Big Data Helps Guiyang Intelligent Transportation. *Computers & Internet*, 2015 (19): 9.
[24] Tian Qiang. Highlight the Advantages of Big Data Application and Innovate the Three-Dimensional Prevention and Control System. *People's Public Security Daily, Traffic Safety Weekly*, October 9, 2015, p. 3.
[25] Hu Caiyi and Yang Xinmiao. Hub Information Service of Integrated Transportation Based on Big Data. *Comprehensive Transportation*, 2015 (07): 60–62.
[26] Bie Kun. Big Data Drives Intelligent Transportation. *Computerworld*, July 15, 2013, p. A10.

[27] Gao Shudong. Challenges Faced by Commercial Banks in the Era of Big Data — Interview with Zhou Yanti, Deputy General Manager of Data Center of Bank of Communications Co., Ltd. *Financial Computer of China*, 2013 (07): 22–24.

[28] Yue Jianming and Yuan Lunqu. Big Data Analysis in the Development of Intelligent Transportation. *Productivity Research*, 2013 (06): 137–138, 165.

[29] Meng Qingfeng. Traffic Problem Solving under Big Data. *China Communications News*, July 17, 2013, p. 5.

[30] Guo Tao. Big Data All in One Machine Makes Urban Traffic Intelligent. *China Information World*, December 31, 2012, p. 16.

[31] Chen Mei. Application of Big Data in Public Transportation. *Library and Information*, 2012 (06): 22–28.

[32] Liu Haiyong. Help Intelligent Growth and Cultivate High-end Talents in the Era of Big Data — IBM and Beijing Jiaotong University Jointly Release the Joint Talent Training Plan for Information Management. *China Education Info*, 2012 (19): 89.

Index

A
abnormal data, 69
accident detector, 105
accidental circumstances, 133
accumulate data and forecast other surrounding areas, 198
accumulation of the big data, 220
accuracy of market demand forecast, 116
accurate and flexible scheduling, 188
active parking system, 164
active users, 63
adaptive cruise control system, 264
ADAS, 243
administrative units, 59
advantages of aggregation, 212
advantages of free travel, 277
afterwards summary and filing, 173
agricultural product information center, 128
airline partners, 34
alarm, 75
Alibaba's big data technology, 228
all data monitoring, 200
all in one network, 56
all round management, 195
Amap, 22
Amazon, ix
analysis by the Los Angeles Institute, 170
analysis of path, 263
analyzing the regional distribution, 113
anti-lock braking system (ABS), 272
application accuracy rate, 133
application data, 57
areas with different road network densities, 274
artificial intelligence technology, 16
Audi, 259
Audi's development plan, 260
authoritative data, 36
automated driving into six levels, 250
automated driving tools, 282
automatic alarm device, 178
automatic driving safety, 282
automatic linkage of vehicles, 298
automatic mode, 93
automatic parking system, 264

automatically perform backward positioning, 300
automation at single-function level, 239
automobile, 251
automotive giants, 54
autonomous driving, 236
autopilot program, 259
autopilot technology, 218
aviation logistics management, 145
aviation logistics, 143–144

B

B2C operation model, 67
back road of manned driving, 286
Baidu group, 8
Baidu group's direct forays, 228
Barcelona, 47
beforehand supervision and prediction, 173
BeiDou system, 85
Beijing bus group, 166
Beijing Olympic Games, 177
Beijing, 82
Beijing's highway road administration system, 30
Beijing's Ring Road, 181
bicycle operating companies, 63
big city problems, 17
big data accurate prediction, 188
big data application of smartphones, 203
big data intelligent equipment, 174
big data platform of the Expo, 184
big data report on intelligent travel, 188
big data resource pool, 14
big data routing calculation, 133
big data trans-nation logistics system, 142
big data traffic route planning of Zigong, 156
big data transportation management model of Hangzhou, 213
big data transportation management models, 210
big data transportation, 3
bin-location of commodities, 125
binary information, 3
black car, 292
black list, 37
blind spots of public transport services, 67
blind traffic mode, 71
blockchain, 7
blossoming state of affairs, 197
blowout development, 55
Bluetooth, 89
Bluetooth-entering, 22
BOLLORE group, 62
booking process, 34
border effect, 191
break the barriers, 275
Bureau of Civil Aviation, 37
bus App, 190
bus card, 202
bus planning, 168
bus priority control system, 180
business chain of enterprises, 148

C

3rd China County E-commerce Summit, 126
Cai Niao network, 126
California Driverless Road Test Permits, 249
California unmanned driving licenses, 252

California, 249
Caoniao logistics, ix
capacity of motorways, 273
capacity supplement, 148
car as a life, 292
car as a service, 292
car driving controller, 269
caravans, 289
cardholder, 82
Carnegie Mellon, 253
carpooling mode, 65
carriageway, 263
car–road, 285
car's reaction time, 274
central nerve, 138
changes of urban traffic flow, 86
characteristic of big data, 155
characteristic of Internet vehicle, 219
China Railway Corporation, 25
China TransInfo Technology Corporation, 225
China's car ownership, 98
China's development process, 41
China's high-speed rail network, 24
China's transportation industry, ix
China's urban logistics, 128
Chinese characteristics, xi
Chongqing north railway station, 108
Chongqing's first comprehensive traffic law enforcement closed management system, 108
City Council, 49
city governance, 9
Civil Aviation Department, 6
civil aviation, 26
civilized tourism, 37
classification of vehicles into different uses, 283
clear operation track, 129
closed-loop effect, 155
cloud accounting platform, 150
cloud computing, 7
cloud emergency system, 119
cloud platform, 14
cloud service layer, 208
cold chain logistics, 134
cold chain products, 136
cold chain transportation, 135
collaboration and interaction, 281
collaborative development, 275
collection effect, 82
command center at all levels, 77
commercial big data analysis tools, 117
commercial vehicles, 238
commercialization process of the unmanned driving technology, 282
common connection among vehicles, 88
common value, 38
communication-based train control, 297
competitive barriers to the development of the unmanned driving cars, 257
complete report, 195
comprehensive travel information service system, 184
conditional automation, 240
conditions to drive, 270
Confusing Road Function, 154
congested road, 275
connection between different subsystems, 296
construction of intelligent transportation in Changsha, 10
construction of urban rail transit system and comprehensive ticketing system, 229

construction of urban traffic accident detection platform, 105
continuous innovation of technology, 203
contribution to the development of human civilization, 287
contributor to air pollution, 248
control center, 302
control module, 273
convenience of ETC system installation and payment, 231
convenient and green, 27
cooperative relationships, 215
coordinated data, 56
core competitiveness, 143
core module, 269
core value, 62
cost of enterprise operation, 115
courier industry, 256
crash safety system, 286
crisis, 132
cross-border logistics, 140
crowd effect, 167
crowdsourcing, 66
current market, 19
customers, 22
customer's image, 34
customized bus, 67, 165
customized exclusive service, 217
customized public transport, 67
customized services for every customer, 151
Cyber C3 smart car, 254
cycle patterns, 88

D

Dada Bus, 66
daily average order of Jingdong, 149
daily management of taxi and bus, 87
daily toll collection, 98
daily vehicle access records, 99
data acquisition, 99, 209
data algorithm for accident detection, 106
data analysis evolution process, 45
data analysis of OFO platform, 21
data analysis, 197, 209
data analytics, 64
data cage, 67
data collection of agricultural products, 135
data cube, 17
data distribution law, 137
data exchange platform, 101
data fragmentation, 141
data from upstream detector, 107
data information protection, 117
data thinking, 210
data tools, 43
data-driven system, 262
decision-making and simulation, 12
decision-making module, 273
declaration on the creation of a State-of-the-Art IT Nation, 51
deep-rooted problems of the industry, 57
Delphi Pike Electric, 278
departments and institutions + third-party enterprises, 214
departments in Zhengzhou, 138
design of road space and landscape, 161
designers, 162
detailed traffic dynamics, 97
detection equipment, 75
developed countries, x

development of human society, xiii
development of urban construction, 199
development philosophy, xi
Didi hitchhiking service, 65
Didi Research Institute, 278
Didi Taxi, 65
different fault structure trees, 74
difficult travel within two kilometers, 20
digital form, 3
digitization, 4
diplomatic event, 191
discovery, 23
domestic and foreign cases, xiv
Donald Shoup's study, 247
downstream detector, 107
Driverless: Intelligent Cars and the Road Ahead, 270
drivers, 22
driving perception technology, 204
driving range, 283
driving route, 95
driving routes, 58
drone delivery of JD, 124
Dt Dream, 187
dual system, 287
Dunda Delivery, 129
duplication of construction, 50
dynamic data, 83
dynamic information collection equipment, 196

E

eagle eye, 97
e-commerce industry, 131
e-commerce merchants, 134
e-waybill, 132
early stage of road design, 162
ecology, 245
economic benefits of logistics service providers and users, 115
effect of synchronous on collection, analysis and suggestion, 174
efficiency and comfort of the trains, 296
efficiency of information utilization, 10
elderly and the disabled people, 257
electronic charging mode, 178
electronic fence, 69
electronic navigation map, 94
electronic navigation, 94
electronic police, 181
electronic toll collection system, 185
electronic toll collection, 97
electronic tolling system, 32
electronics giants, 54
emergence of the information technology revolution, 3
emergency accident handling in operation, 299
emergency accident rescue, 52
emergency braking and satellite navigation system, 264
emergency command platform, 118, 120
emergency command service platform, 173
emergency rescue system, 33
emergency rescue, 298
emergency signals on the road, 93
emergency support articles, 175
emergency system, 172
emerging countries, 50
emerging IT enterprises, 205
energy convenience, 290
enrichment of data, 200

enterprise production and operation, 115
entrance, 81
environment of the road, 162
environmentally friendly route, 53
ETC lane coverage rate, 98
ETC network of expressway, 6
ETC system, 100
European Union (EU), 47
evacuation platform, 301
excessive traffic flow, 199
exchanging vehicles, 284
experimental and testing stage, 237
exploration of traffic management models, 275
exploration stage, 167
express delivery system, 132
express logistics, 131
expressway management platform, 104
extension of the living space, 289
extensive application of ETC system in the transportation fields, 231

F

13th Five-Year Plan, 229
fault analysis, 74
false orders, 134
Federal Big Data Research and Development Strategic Plan, 42
Fengchao self-delivery, 131
Fiberhome FitData, 74
field of transportation, 191
final data, 207
final goal for the intelligent automobile, 244
financial investment, 59
first batch, 232
first draft of the Management Specification for the Adaptability Verification of Intelligent Connected Vehicles on Public Roads, 5
first generation, 100
first intelligent tourism highway in Jiangxi Province, 102
first ITS World Congress, 7
first smart car in 1989, 254
first-tier cities, 20
five application software systems, 11
five basic management models, 212
five-year partnership with Toyota, 260
fleet management, 45
Flight Manager, 73
floating car data (FCD), 52
focus of public attention, 216
focus of rumors, 216
forecasting, 182
foreign experience, xiv
foreign unmanned driving technology, 294
format problems, 50
four new inventions, 58
four systems, 186
four-level address library, 133
four-stage, 154
fourth largest federal government fleet in the United States, 44
Fourth National GPS Operator Conference, 88
free charging stations, 47
free travel, 276
freight data, 38
freight drivers, 38
freight transportation, 38
freight vehicle, 39

freight volume, 39
full automation, 240
fully automated unmanned driving subway, 293
function of detecting and handling obstacles automatically, 300
function of return prediction, 189
functional features of the unmanned driving systems in urban rail transit, 295
funding, 46
future road network structure and space allocation model of Xiong'an New District, 8

G

geographical limitations, 186
G20 summit, 186
gate data, 111
geographic data information systems, 95
geographic data, 43
geospatial data processing, 47
Germany, 49
giants of e-commerce and express industry, 132
global coverage, 96
global penetration, 250
globalization, 140
golden period of rapid development, 47
goods, 39
government agencies, 73
government departments, 20
government emergency management, 118
government framework, 42
government intranet, extranet, and internet, 119
Government of Japan, 51
government traffic management platform, 4
government-guided norms, 42
green effect, 63
green travel, 223
ground induction coil, 87
Guan Che Bao, 39
Guangfo transportation hub, 168
Guangzhou–Foshan Metro, 169
guidance of the Ministry of Public Security and the Public Security Cloud Computing Construction Center, 13
Guiyang city, 201
Guiyang Transportation Bureau, 69
Guiyang, 56
Guiyang's transformation of urban transportation, 202
Guizhou HighTech Zone, 167
Guizhou Internet + road traffic management model, 13
Guizhou province, 12
Guizhou traffic police cloud, 15
Guizhou traffic police, 12

H

Haikou Meilan International Airport, 36
Hangzhou traffic management department, 214
Hangzhou traffic police department, 214
Hangzhou, 105, 186
Hangzhou's traffic, 189
hardware, 89
hazard information provision, 52
high definition digital detection system, 179

high dependence, 301
high safety, 296
high vacancy rate of cars, 65
high-definition digital maps, 267
high-definition digital video monitoring, 76
high-speed development of startups, 252
high-speed railway travel, 26
highway environment, 236
highway management department, 29
highway network monitoring system, 91
Hisense, 17
Hohhot, 164
homemade unmanned driving car, 259
Honda Motor Company's Internavi, 52
hub planning, 166
huge data, 83
human and material resources, 29
human face recognition system, 26
human intelligence, 16
human intervention, 135
human or car, 286
human support, 302
human-caused traffic accidents account, 285
humane and intelligent considerations, 276
human–computer interaction technology, 270
human–machine monitoring interface, 296

I

3D images, 254
IC card, 84
immediate warning, 90
implementation of the unmanned operation, 265
Implementation Plan on Promoting the Internet + Convenient Transportation to Promote the Development of Intelligent Transportation, 233
implementation plan, 233
import and apply, 79
improper transportation capacity, 170
In Aichi Prefecture, 294
increase capital investment, 233
increasing improvement of urban rail transit, 295
increasing popularity of the shared car, 245
independent brands, 218
industry chain, 44
information analysis in real time, 200
information barriers, 212
information infrastructure, 49
information island effect, 207
information island, xi
information layer, 208
information resources, 11
information superhighway, 42
information traceability, 136
innovation and development, 230
innovative cases, 61
Inrix, 225
installation and debugging of the equipment, 174
instructional cars, 69
Insurance Institute for Highway Safety in the United States, 247
integrated classification of goods, 143
integration model, 211
integrity, 159

Intellectual Traffic Safety Association of Shenzhen, 206
intelligent airport, 144
intelligent automobile laws and regulations, 244
intelligent automobile standards, 244
intelligent bus dispatching center, 170
intelligent control systems, 271
intelligent development of China's automobile industry, 5
intelligent digital image information processing technology, 92
intelligent interconnection, 5
intelligent operation and dispatching system of Olympic bus, 179
intelligent prediction system, 194
intelligent quotation and intelligent scheduling, 146
intelligent traffic recognition algorithm, 204
intelligent transportation system, 7, 244, 268
intelligent travel in Nanjing, 172
interaction of each system, 272
interim emergency response and scheduling, 173
international competition, 55
international leading level, xiii
Internavi information center, 53
Internet + action, 230
Internet + convenient transportation, 233
Internet + traffic, x
Internet + transportation, 6
Internet age, xiii
Internet channels, 120
Internet conflicts, 131
Internet of things sensor, 203
Internet of vehicles, 31
Internet skills, 219
intersection of Huangjin Road and Zaoshan Road, 201
investment and financing boom of the big data transportation industry, 226

J

Japanese companies, 55
Japan Meteorological Society's database, 54
Jiaxing, 165
Jinan municipal government, 75
Jiuqu Data, 227
journey of the development of the unmanned driving technology, 238

K

key to automated driving, 269
key driver credit system, 14

L

labor-saving, 153
lack of data security, 60
lack of national unified planning, 58
lack of surrounding parking lots, 199
lane deviation detection system, 264
lane separation, 284
large-scale activities, 175
later stage of design, 163
law enforcement behavior of law enforcement personnel, 112
law enforcement station and service areas, 15
leader of the big data transportation industry, 217
leadership of the government, 130
legality of the unmanned driving, 288
Level 0, 239

Level 1, 239
Level 2, 239
Level 5, 240
Li Yongle's wireless network technology, 267
lidar, 266
life cycle of a vehicle, 46
loading objects, 283
location of the traffic accident, 107
logistics and supermarket distribution, 149
logistics costs, 149
logistics enterprises, 124
logistics experience of the shipper and the carrier, 147
logistics in China, 123
logistics industry management, 118
logistics industry, 117
logistics information platform, 115
logistics transportation channels, 138
logistics transportation, 123
loopholes, 157
low-carbon development of transportation, 81

M

macro-background, xi
Made in China, ix
main mode of urban motorized transport, 279
mainstream domestic express companies, 132
mainstream tool for urban residents' daily travel, 299
mainstream trend of the unmanned driving car exploration projects, 266
maintain efficient coordination, 302
major world events, 177
management measures, 183
management of Guiyang Low Carbon Transportation Information Command Center, 68
management of logistics cost, 150
management of shipping companies, 145
managerial confusion, 59
manual handling, 301
manual mode, 93
manufacturing and operations of the unmanned vehicles, 252
map software, 21
market operation, 230
market scale of China's satellite navigation products and services, 224
market-led development, 42
mass carpooling, 280
mass storage and rapid calculation, 113
massive data, 11
mature technology applications, 99
medium and long-term Development Plan of the National Satellite Navigation Industry, 224
Metro Line 1 in Paris, 294
metro lines in Guangzhou, 168
metro operating companies, 24
metro operations, 299
metro planning, 168
micro data collection, 200
microcirculation pilot, 160
Microsoft, 260
millimeter-wave radar, 266
mine and meet the demands, 70
mixed data, 158
mobile APP, 6
mobile charging, 290

mobile handling plan, 180
mobile patrol car, 181
mobile phone for transportation services, 194
mobile phone map, 4
mobile platform, 120
Mobileye, 258
modern logistics and traditional logistics, 125
monitoring equipment, 120
monitoring system, 102, 179
monitoring the regional video, 113
more obvious advantages, 27
most advanced toll collection method, 219
movement mode, 87
multi parameter and large-scale detection, 93
multi-day travel, 82
multi-platform and multi-channel travel mode, 167
multi-point cooperation and all-round development, 210
multi-stage detection, 94
multimodal transportation, 137
Municipal Transportation Commission, 56
MySQL database, 201

N

Nanjing Metro, 22
Nanning Rail Transit APP, 23
national conditions, xiv
National Convention Center of Chaoyang District, 193
national economic development, 186
national economy, x
National Highway Traffic Safety Administration (NHTSA), 239
National Information Center, 222
national intelligent bus card, 84
national policy field, 229
national road network, 91
national strategy, 51
National Tourism Administration, 37
navigation services, 73
navigation industry, 223
neighboring countries, 141
network cable planning, 158
network communication layer, 208
network in Beijing, 178
network line congestion, 157
network line planning, 157
network-based database, 101
new energy electric vehicle, 223
New Silk Road Challenge plan, 255
new traffic characteristics of Beijing, 193
new urban rail transit technology, 293
new way of thinking, xi
NFC, 84
Ningdu–Dingnan expressway, 101
no automation, 239
no reasonable network, 154
nodes and multi-technologies, 111
non-paid area, 110
non-popularity, 220
non-truck operating common carrier, 145
normal operation, 297
number of passengers, 65, 283
number of private vehicles, 247
number of share on the lane, 108

O

OBD, 89
OBD2 system, 54
objectivity and calmness, 271

obtain the function of video monitoring, 300
occupancy rate, 106
odd-and-even license plate rule, 180
official platform and certification platform, 215
on-board sensors, 273
onboard electronic navigation map industry, 94
one machine in hand, free travel in the world, 190
one shared SAV, 256
one stop of information services, 138
one website, one phone, 184
One-Belt-One-Road Forum, 191
one-day travel, 82
online testing and monitoring, 298
open government, 43
open-loop, 155
operating platforms, 83
operational pressure of express companies, 133
optical fiber, 103
optimal driving path, 268
optimization of resource allocation, 213
orderly construction of intelligent application demonstration projects of urban public transportation, 232
organization model, 207
organization, 123
original technical and management obstacles, 151
Oslo, 279
outbreak of industry, 229
overall management awareness of chess, 10
overseas warehouse, 142

P

package volume skyrocketed, 134
pace of development, 72
paid area, 110
Paris electric car rental system, 62
parking guidance system, 164
partial unmanned driving car research and development, 261
parts suppliers, 251
passenger flow detection platform, 108
passenger flow detection platform, 110
passengers' interference, 303
path planning technology, 263
path planning, 263
peak of large passenger flow, 109
people-oriented development, 280
people's travel methods, 19, 278
people's travel, 41
per capita vehicle ownership, 44
perception level, 265
perfect emergency planning scheme, 172
personal information, 34
personalized service, 220
pilot cities, 232
plan the driving routes, 267
planning and design of public transport, 108
platform area, 110
police affairs cloud big data processing technology, 76
position in the global automobile development, 238
positioning and navigation system, 262
precise data, 66
prediction of the industry authorities, 246

preliminary research, 161
primary consideration for an unmanned driving car, 269
priority signal control points, 76
private price hike of taxis, 221
problem of "last kilometer" delivery, 126
process of autopilot, 286
process of online shopping, 134
process of transportation law enforcement, 111
processing, 198
product of the development of various technologies, 243
product quality, 139
progress of the automobile industry, 238
promoting traffic big data, x
provide real-time navigation services, 267
public bicycles, 20
public misunderstanding and mistrust of the big data information, 215
public participation, 162
public platform for road freight vehicles, 148
public security information resources, 14
public transportation system, 32
public transportation, 22
public's online complaints and reports, 114
Pudong Airport, 35
Pudong Punctuality, 144

Q

Qinglong platform, 115
quality shared mobility services at a lower cost, 279
quantitative information, 147
quest for safety, 285

R

Rail Transit Big Data Expert System, 74
R&D stage, 218
railway supply chain, 138
rapid development momentum, 217
rapid passenger transport network, 25
raw data, 43
RBMS flow meter, 178
reasons of traffic jam, 199
regional adaptive control system, 77
regional meteorological data and information, 104
relatives' travel, 73
relevant policies and norms, 50
relevant service enterprises, 58
remote control system, 139
research on the unmanned driving cars in Western developed countries, 253
researchers at the University of Texas at Austin, 248
resource allocation, 62
resource sharing and department cooperation, 213
response plans, 83
restricted routes, 192
RFID technology, 142
risk and vulnerability analysis, 211
road administration bureau, 49
road administration management, 30
Road Condition Detection System, 93
road information, 29
road network maintenance and management, 29

roads that are highly prone to congestion, 275
root of the problem, 300
route planning and network line planning, 160
route planning, 153
rural logistics professionals, 128
rural logistics, 126
rural version of the Internet car hailing mode, 221
rush hour, 100

S

2018 Spring Festival Transportation Big Data Report, 25
2018 Spring Festival transportation, 33
1 + 6 system, 206
safe and standardized application and development of the unmanned driving technology, 303
safety accidents, 301
safety and good airport management, 34
safety management, 29
satellite map data, 203
saves the travel time, 27
science and technology, 206
scientific decisions, 92
second batch, 232
security of the logistics information platform, 117
seizing the wave of informatization, 4
self-learning and self-improving capabilities, 271
self-processing, 69
self-service bicycle travel, 20
self-service boarding, 36
self-service luggage check-in, 35
self-service security check, 35
self-service travel, 21
semi-automated ones, 242
semi-autonomous and fully automated vehicles, 246
sensor data, 57
sensors, 49
series of basic and business databases, 103
service management, 30
service quality, 36
several characteristics, 61
several kinds of highway travel, 28
Shandong province, 17
Shanghai Jing'an Branch, 112
Shanghai Metro Line 10, 295
Shanghai transportation big data, 182
Shanghai transportation department, 183
Shanghai World Expo, 182
Shanghai's network line planning, 159
shared culture, 222
shared economy model, 61
shared transportation culture, 222
shared travel, 277
shared warehouse cross-border ecological service platform, 140
sharing delivery, 129
sharing models, 64
Shenyang, 164
Shenzhen's model, 205
shipper, 39
short distance travel, 28
short distance, 21
short message technology, 86
short-distance and long-distance vehicles, 291

Shunfeng, 131
signal, 301
Singapore Automated Road Traffic Committee, 276
Singapore, 278
situation of chaos after large-scale activities, 189
six systems, 11
skeleton for the automotive system, 262
Sky Network + Ground Network + People Network, ix
smaller vehicles, 280
smart advertising screen, 48
smart bus shelters, 48
smart city development, 8
smart earth, 18
smart grid roadmap, 250
smart lock, 63
smart quick pay system, 48
smart safety, 259
smart street, 163
smart tires, 225
smart touch system, 48
smart travel ecology, 281
smartphones, 87
smooth roads, 274
social life and consumption patterns, 115
social resources, 91
social transportation capacity, 127
social vehicles, 129
Society of Motor Vehicle Engineers (SAE), 240
software modules, 103
sorting speed, 133
Southeast Asia smart logistics big data center, 141
spatial data system, 95
spatial state analysis, 211
special blocks for the unmanned driving vehicles, 256
special environments, 237
special intelligent robot, 236
special logistics information platform, 127
special national conditions, 58
special scenes, 283
special service tasks, 175
special transportation capacity ecosystem, 127
specialized processing and value mining of the mastered information, 150
speech recognition technology, 224
stand-alone intelligent vehicle, 268
standard van type freight ETC, 231
standardized container, 139
Stanford University and MIT, 261
start-up companies, 228
State Council, 230
state diagnosis of transportation, 195
state diagnosis, 196, 198
static + dynamic, 197
static data, 83
static information collection, 196
station diversion, 165
station planning, 163, 165
stations, 302
statistical bulletin on the Development of the Transport Industry, 38
statistics department, 82
statistics of each bus information, 165
strategic cooperation, 226
strategy of balanced entry into the park, 183

streamline, 110
strengthening the operation of big data, 121
subversion of traditional manual management methods, 297
subway lines in China, 294
summary analysis report, 46
summit forum, 192
support of the unmanned driving technology, 245
survey by RAND Corporation, 246
survey by the Virginia Tech Transportation Institute (VTTI), 242
system comprehensive analysis model, 212
system construction of credit evaluation system business, 12
system data, 57
system failure, 298
system intelligent collection and artificial screening, 112

T

2035, 246
T-union, 81
tachograph, 225
tailored taxi service, 67
taxi startup nuTonomy, 278
taxi trajectory data, 86
tech giants, 251
technical difficulties, 299
technologies used in the environmental sensing, 267
technology of traffic management, 202
temporary transportation capacity, 128
Teradata Dynamic Enterprise Data Warehouse Platform 6690, 45
Teradata geospatial feature, 46
Tesla, 258
The Barcelona Metro Line 9 in Spain, 294
the Internet and traditional industries, 6
thinking mode of transportation management in the past, 10
third-party big data companies, 188
third-party platform, 104
Thousand Talents Plan, 227
three major potential needs, 244
three technological changes, 204
three-dimensional transportation information system, 12
three-layer structure, 208
three-way risk control system, 147
thriving scientific technology, 235
ticket price inquiry, 23
tidal effect, 161
timeliness guarantee, 137
TOCC system, 194
Tongcuncun APP, 221
top 100 Best Fleets, 47
top automobile brands in the world, 226
Toyota Motor Corporation, 253
Toyota Research Institute (TRI), 260
Traction or Stability Control Systems, 272
trade country, 123
traditional automobile manufacturers, 251
traditional road planning, 160
traditional traffic route planning, 154
traditional transportation, 9
traditional vehicles, 89
traffic accidents, 105
traffic brain, 156

traffic conditions, 72
traffic congestion, 9
traffic deaths, 51
traffic decision system, 185
traffic departments, 288
traffic flow detection systems, 92
traffic flow detector, 106
traffic flow, 75
traffic in Shenzhen, 205
traffic inconvenience, 168
traffic information in real time, 71
traffic issues, 187
traffic Management Bureau of the Ministry of Public Security, 16
traffic management department + third-party big data enterprise, 209
traffic management information, 31
traffic managers, 65
traffic map special data, 96
traffic planning in major activities, 177
traffic pollution, 41
traffic queues, 185
traffic restricted, 192
traffic security integrated information command center of the G20 Summit, 187
traffic signals, 89
traffic supervision organization, 96
traffic volume, 104
traffic zone controller, 30
train accident, 303
train carpooling, 66
training of the drivers, 69
trajectory tracking model, 211
transaction safety, 146
transit metropolis, 159
transport status of transport vehicles, 116
transportation Big Data Center, 56
transportation capacity, 169
transportation community, 190
transportation distribution management, 125
transportation ecology, 245
transportation efficiency, 40
transportation industry, 4
transportation information resources, 10
transportation information system, 31
transportation management, 5
transportation sector, 44
transportation speed, 139
transportation system, 24
travel behavior and habits, 19
travel efficiency of residents, 90
travel on separate roads, 284
traveler's behavior data, 24
TravelSky's Umetrip Software, 36
trial project, 68
trust from the public, 216
two city interaction situation, 167
two networks and one platform, 196
two platforms of intelligent transportation management, 11
two-way infrared technology, 110

U

UAV transportation service, 124
U.S. Air Force, 44
U.S. GPS satellite, 85
Uber, 248
uncivilized air travel behavior, 37
unforeseen situation, 299
unified control of distribution vehicles, 129

unified cross-border logistics big data platform, 141
unified data sharing platform, 60
unified law enforcement platform, 112
unified sharing standard, 102
University City, 68
unmanned car FAW Hongqi HQ3, 255
unmanned driving automobile, 236
Unmanned Driving Car Industry Development Prospects and Investment Strategic Planning Analysis Report, 261
unmanned driving car Lux, 253
unmanned driving car manufacturers, 291
unmanned driving conditions, 291
unmanned driving environment, 292
unmanned driving intelligent car project, 255
unmanned driving legislation, 249
unmanned driving Saikabo, 254
unmanned driving taxis, 277
unmanned driving trains, 293
unmanned private cars, 8
urban and rural passenger transportation, 171
urban delivery, 130
urban development, 153
urban environments, 237
urban intelligent transportation system, 17
urban logistics storage platform, 130
urban operation monitoring data, 119
urban traffic congestion, 11
urban traffic, 18
US Highway Traffic Safety Administration, 249
user demand, 70
user experience, 62, 147
users' demand for unmanned cars, 289
users' privacy, 292
Uterra, 254

V

various factors which may affect the traffic flow, 273
various means of transportation, 19
vehicle body, 289
vehicle collisions, 158
vehicle control system, 32
vehicle control technology, 264
vehicle data, 90
vehicle identity and authentication system, 136
Vehicle Information and Communication System (VICS), 52
vehicle operating zone, 283
vehicle speed, 108
vehicle-road collaboration technology, 218
vehicle-to-vehicle docking, 290
vehicles with special certificates, 193
vehicle's internal and external interfaces, 288
Velodyne LiDAR, 281
VICS real-time information, 53
video information of key areas, 113
video monitoring system, 71
visual telephone, 30
Volvo, 258

W

waste of resources, 50
waste of traffic resources, 170
WeChat service account, 77

Weifang Transportation Emergency Command Center, 118
well-developed unmanned driving system, 265
wheeled mobile robot, 243
Wi-Fi, 48
wireless operators, 87
workflow, 45
World Economic Forum, 246
World Health Organization, 285
world's big data center, 43
world's first alternating current generator, 287
world's first UAV operation and dispatching center, 124
world's longest total length of high-speed railway and the highest frequency of travel for the public, 27
Wuhan Traffic Management Bureau, 85
Wuxi Research Institute, 16

X

Xiong'an New District, 8
Xizhimen hub, 166
Xizhimen transportation hub, 169

Y

Yuantiao Technology, 227
Yunnan Intelligent Transportation, 146
Yuyao bus company, 171
Yuyao transportation department, 171
Yuyao, 171

Z

Zebra technology, 218
Zero Private Car, 279
Zhonghe technology, 229
Zigong, 156

Printed in the United States
by Baker & Taylor Publisher Services